Even You Can Learn Statistics and Analytics

Fourth Edition

An Easy to Understand Guide to Statistics and Analytics

David M. Levine

David F. Stephan

Boston • Columbus • New York • San Francisco • Amsterdam • Cape Town
Dubai • London • Madrid • Milan • Munich • Paris • Montreal • Toronto • Delhi • Mexico City
São Paulo • Sidney • Hong Kong • Seoul • Singapore • Taipei • Tokyo

Editor-in-Chief: Mark L. Taub
Acquisitions Editor: Kim Spenceley
Development Editor: Chris Zahn
Managing Editor: Sandra Schroeder
Project Editor: Mandie Frank
Production Manager: Remya Divakaran/codeMantra
Copy Editor: Kitty Wilson
Indexer: Timothy Wright
Proofreader: Donna Mulder
Designer: Chuti Prasertsith
Compositor: codeMantra

1 2022

ISBN-13: 978-013-765476-5
ISBN-10: 0-13-765476-6

Library of Congress Control Number: 2021947626

Pearson's Commitment to Diversity, Equity, and Inclusion

Pearson is dedicated to creating bias-free content that reflects the diversity of all learners. We embrace the many dimensions of diversity, including but not limited to race, ethnicity, gender, socioeconomic status, ability, age, sexual orientation, and religious or political beliefs.

Education is a powerful force for equity and change in our world. It has the potential to deliver opportunities that improve lives and enable economic mobility. As we work with authors to create content for every product and service, we acknowledge our responsibility to demonstrate inclusivity and incorporate diverse scholarship so that everyone can achieve their potential through learning. As the world's leading learning company, we have a duty to help drive change and live up to our purpose to help more people create a better life for themselves and to create a better world.

Our ambition is to purposefully contribute to a world where:

- Everyone has an equitable and lifelong opportunity to succeed through learning.
- Our educational products and services are inclusive and represent the rich diversity of learners.
- Our educational content accurately reflects the histories and experiences of the learners we serve.
- Our educational content prompts deeper discussions with learners and motivates them to expand their own learning (and worldview).

While we work hard to present unbiased content, we want to hear from you about any concerns or needs with this Pearson product so that we can investigate and address them.

- Please contact us with concerns about any potential bias at https://www.pearson.com/report-bias.html.

Credits

Cover	ZinetroN/Shutterstock
Unnumbered Figure 3-1 – Unnumbered Figure 3-3	Microsoft Corporation
Unnumbered Figure 5-1 – Unnumbered Figure 5-3	
Figure 6-2	
Figure 6-3	
Figure 8-2 – Figure 8-5	
Unnumbered Figure 8-1	
Unnumbered Figure 8-2	
Figure 9-1 – Figure 9-3	
Figure 9-5	
Figure 9-6	
Figure 10-3	
Figure 11-1	
Figure 12-1 – Figure 12-3	
Figure 12-5 – Figure 12-7	
Figure E-1 – Figure E-5	
Unnumbered Figure E-1	
Unnumbered Figure E-2	
Figure 13-5	JMP Statistical Discovery LLC
Figure 13-6	

To our wives and our children,
and in loving memory of our parents

Table of Contents

Acknowledgments

We would especially like to thank the staff at Pearson: Kim Spenceley for making this fourth edition a reality, Kitty Wilson for her copy editing, Lori Lyons and Mandie Frank for their work in the production of this text.

We have sought to make the contents of this book as clear, accurate, and error-free as possible. We invite you to make suggestions or ask questions about the content if you think we have fallen short of our goals in any way. Please email your comments to authors@davidlevinestatistics.com and include the hashtag #EYCLSA4 in the subject line of your message.

About the Authors

David M. Levine and **David F. Stephan** are part of a writing team known for their series of business statistics textbooks that include *Basic Business Statistics, Business Statistics: A First Course*, and *Statistics for Managers Using Microsoft Excel*. In long teaching careers at Baruch College, both were known for their classroom innovations, with Levine being honored with a Presidential Excellence Award for Distinguished Teaching Award and Stephan granted the privilege to design and develop the College's first computer-based classroom. Both are active members of the Data, Analytics and Statistics Instruction SIG of the Decision Sciences Institute.

Levine is Professor Emeritus of Information Systems at Baruch College. He is nationally recognized innovator in business statistics education and is also the coauthor of *Applied Statistics for Engineers and Scientists Using Microsoft Excel and Minitab*. Levine is also the author or coauthor of four books about statistical quality management: *Statistics for Six Sigma Green Belts and Champions, Six Sigma for Green Belts and Champions, Design for Six Sigma for Green Belts and Champions*, and *Quality Management*, 3rd Edition. He has published articles in various journals, including *Psychometrika, The American Statistician, Communications in Statistics, Multivariate Behavioral Research, Journal of Systems Management, Quality Progress,* and *The American Anthropologist*, and has given numerous talks at American Statistical Association, Decision Sciences Institute, and Making Statistics More Effective in Schools of Business conferences.

During his more than 20 years at Baruch College, **Stephan** devised techniques for teaching computer applications such as Microsoft Excel in a business context and developed future-forward courses that explored the effects of emerging digital technologies. He also served as the associate director of a U.S. Department of Education FIPSE project that successfully integrated interactive media into classroom instruction for the humanities.. Stephan is also the developer of PHStat, the statistics add-in for Microsoft Excel distributed by Pearson Education.

Introduction
The *Even You Can Learn Statistics and Analytics* Owner's Manual

In today's world, understanding statistics and analytics is more important than ever before. *Even You Can Learn Statistics and Analytics: An Easy to Understand Guide to Statistics and Analytics* teaches you the basic concepts that provide you with the knowledge to apply statistics and analytics in your life. You will also learn the most commonly used statistical methods and have the opportunity to practice those methods while using Microsoft Excel.

Please read the rest of this introduction so that you can become familiar with the distinctive features of this book. To download files that support your learning of statistics, visit the website for this book at www.informit.com.

Mathematics Is Always Optional!

Never mastered higher mathematics—or generally fearful of math? Not to worry, because in *Even You Can Learn Statistics and Analytics*, you will find that every concept is explained in plain English, without the use of higher mathematics or mathematical symbols. However, if you *are* interested in the mathematical foundations behind statistics, *Even You Can Learn Statistics and Analytics* includes **Equation Blackboards**, stand-alone sections that present the equations behind statistical methods and complement the main material.

Learning with the Concept-Interpretation Approach

Even You Can Learn Statistics and Analytics uses a **Concept-Interpretation** approach to help you learn statistics and analytics:

- A **CONCEPT**, a plain language definition that uses no complicated mathematical terms.
- An **INTERPRETATION**, that fully explains the concept and its importance to statistics. When necessary, these sections also include common misconceptions about the concept as well as the common errors people can make when trying to apply the concept.

For simpler concepts, an **EXAMPLES** section lists real-life examples or applications of the statistical concepts. For more involved concepts, **WORKED-OUT PROBLEMS** provide complete solutions to statistical problems—including actual spreadsheet results—that illustrate how you can apply the concepts to other problems.

Practicing Statistics While You Learn Statistics

To help you learn statistics, you should always review the worked-out problems that appear in this book. As you review them, you can practice what you have just learned by using the optional **SPREADSHEET SOLUTION** sections.

Spreadsheet Solution sections enable you to use Microsoft Excel as you learn statistics. If you don't want to practice your spreadsheet skills, you can examine the spreadsheet results that appear throughout the book. Many spreadsheet results are available as files that you can download for free through the InformIT website, www.informit.com. Please visit the website for this book at www.informit.com to access these bonus materials.

Spreadsheet program users will also benefit from Appendix D and Appendix E, which help teach you more about spreadsheets as you learn statistics.

And if technical issues or instructions have ever confounded your using Microsoft Excel in the past, check out Appendix A, which details the technical configuration issues you might face and explains the conventions used in all technical instructions that appear in this book.

In-Chapter Aids

As you read a chapter, look for the following icons for extra help:

Important Point icons highlight key definitions and explanations.

File icons identify the downloadable files that enable you to examine the data in selected problems.

Interested in the mathematical foundations of statistics? Then look for the Interested in Math? icons throughout the book. But remember, you can skip any or all of the math sections without losing any comprehension of the statistical methods presented, because math is always optional in this book!

End-of-Chapter Features

At the end of most chapters of *Even You Can Learn Statistics and Analytics,* you can find the following features, which you can review to reinforce your learning.

Important Equations

The **Important Equations** sections present all of the important equations discussed in the chapter. You can use these lists for reference and later study even if you have skipped over the Equation Blackboards and "interested in math" passages.

One-Minute Summaries

Each **One-Minute Summary** is a quick review of the significant topics in the chapter in outline form. When appropriate, the summaries also help guide you to make the right decisions about applying statistics to the data you seek to analyze.

Test Yourself

The **Test Yourself** sections offer a set of short-answer questions and problems that enable you to review and test yourself (with answers provided) to see how much you have retained of the concepts presented in a chapter.

Summary

Even You Can Learn Statistics and Analytics can help you whether you are taking a formal course in data analysis, brushing up on your knowledge of statistics for a specific analysis, or need to learn about analytics. If you have questions about this book, feel free to contact the authors via email at authors@davidlevinestatistics. com and include the hashtag #EYCLSA4 in the subject line of your email.

1

Fundamentals of Statistics

Every day, people use numbers to describe or analyze our world:

- **4 out of 5 people don't want their personal data collected or shared without consent.** Invisibly.com reports that in a survey of 1,247 people, 82% of respondents supported measures that would prevent companies and devices from collecting or sharing their data, and 68% of respondents said that data privacy is important to them.

- **Learn more, earn more: Education leads to higher wages, lower unemployment.** A 2020 U.S. Bureau of Labor Statistics report noted that workers aged 25 and over who have less education than a high school diploma had the highest unemployment rate (5.4%) and lowest median weekly earnings ($592) in 2019. Workers with graduate degrees had the lowest unemployment rates and highest earnings.

- **Streaming media device market to hit USD 24 billion by 2026.** Market Research Future, a leading market research firm, expects the size of the streaming media device market to grow to $24 billion by 2026, growing at a compound annual growth rate of 17.6% from the 2020 market size.

You can make better sense of the numbers you encounter if you learn to understand statistics. **Statistics**, a branch of mathematics, uses procedures that enable you to correctly analyze the numbers. These procedures, or **statistical methods**, transform numbers into useful information that you can use when making decisions about the numbers. Statistical methods can also tell you the known risks associated with making a decision as well as help you make more consistent judgments about the numbers.

Learning statistics or analytics requires you to reflect on the significance and the importance of the results to the decision-making process you face. This statistical interpretation means knowing when to ignore results because they are misleading, are produced by incorrect methods, or just restate the obvious, as in "100% of the authors of this book are named 'David.'"

In this chapter, you begin by learning five basic words—*population*, *sample*, *variable*, *parameter*, and *statistic* (singular)—that identify the fundamental concepts of statistics. These five words, and the other concepts that this chapter introduces, help you understand the statistical methods that later chapters discuss.

1.1   The First Three Words of Statistics

You've already learned that statistics is about analyzing things. Although *numbers* was the word used to represent things in the opening of this chapter, the first three words of statistics, *population*, *sample*, and *variable*, help you to better identify what you analyze with statistics.

Population

CONCEPT   All the members of a group about which you want to reach a conclusion.

EXAMPLES   All U.S. citizens who are currently registered to vote, all patients treated at a particular hospital last year, the entire set of individuals who accessed a website on a particular day.

Sample

CONCEPT   The part of the population selected for analysis.

EXAMPLES   The registered voters selected to participate in a recent survey concerning their intention to vote in the next election, the patients selected to fill out a patient satisfaction questionnaire, 100 boxes of cereal selected from a factory's production line, 500 individuals who accessed a website on a particular day.

Variable

CONCEPT   A characteristic of an item or an individual that will be analyzed using statistics.

EXAMPLES Age, the party affiliation of a registered voter, the household income of the citizens who live in a specific geographical area, the publishing category of a book (hardcover, trade paperback, mass-market paperback, textbook), the number of cell phones in a household.

INTERPRETATION Although people often say that they are analyzing their data, they are, more precisely, analyzing their variables. Variables are either **categorical**—variables that contain non-numerical data, data not intended for mathematical calculations—or **numerical**, variables whose data represent a counted or measured quantity. The following table presents more information about the two types of variables, including the two subtypes of numerical variables.

	Categorical Variables	**Numerical Variables**
Concept	The values of these variables are selected from an established list of categories.	The values of these variables involve a counted or measured value.
Subtypes	None	**Discrete** values are counts of things. **Continuous** values are measures, and any value can theoretically occur, limited only by the precision of the measuring process.
Examples	Wears glasses, a variable that has the categories "yes" and "no." Academic major, a variable that might have the categories "English," "Math," "Science," and "History," among others.	The number of people living in a household, a discrete numerical variable. The time it takes for someone to commute to work, a continuous numerical variable.

You should distinguish a variable, such as age, from its value for an individual item, such as 21. An **observation** is the set of values for an individual item in the sample. For example, a sample that contains the variables first name, age, and employed might include the three observations Avery, 33, yes; Jamie, 27, yes; and Peyton, 45, no.

By convention, when you organize data in tabular form, you place the values for a variable to be analyzed in a column. Therefore, some people refer to a variable as a *column of data*. Likewise, some people call an observation a *row of data*.

important point ✎

Every variable should have an **operational definition**, a universally accepted meaning that is understood by all working with the variable. For example, in a previous example the variable *employed* was defined to have yes and no as its values and age was defined as whole years. Without operational definitions, confusion can occur. A famous example of such confusion was a survey that asked about *sex*, to which a number of survey takers answered yes and not male or female, as the survey writer had intended.

1.2   The Fourth and Fifth Words

After you know what you are analyzing, or, using the words of Section 1.1, after you have identified the variables from the population or sample under study, you can define the **parameters** and **statistics** that your analysis will determine.

Parameter

CONCEPT   A numerical measure that describes a variable (characteristic) from a population.

EXAMPLES   The percentage of all registered voters who intend to vote in the next election, the percentage of all patients who are very satisfied with the care they received, the mean time that all visitors spent on a website during a particular day.

Statistic

CONCEPT   A numerical measure that describes a variable (characteristic) of a sample (part of a population).

EXAMPLES   The percentage of registered voters in a sample who intend to vote in the next election, the percentage of patients in a sample who are very satisfied with the care they received, the mean time that a sample of visitors spent on a website during a particular day.

INTERPRETATION   Calculating statistics for a sample is the most common activity because collecting population data is impractical in many actual decision-making situations.

1.3   The Branches of Statistics

You can use parameters and statistics either to describe your variables or to reach conclusions about your data. These two uses define the two branches of statistics: **descriptive statistics** and **inferential statistics**.

Descriptive Statistics

CONCEPT   The branch of statistics that focuses on collecting, summarizing, and presenting a set of data.

EXAMPLES   The mean age of citizens who live in a certain geographical area, the mean length of all books about statistics, the variation in the time that visitors spent visiting a website.

INTERPRETATION   You are most likely to be familiar with this branch of statistics because many examples arise in everyday life. Descriptive statistics serves as the basis for analysis and discussion in fields as diverse as securities trading, the social sciences, government, the health sciences, and professional sports. Descriptive methods can seem deceptively easy to apply because they are often easily accessible in calculating and computing devices. However, this ease does not mean that descriptive methods are without their pitfalls, as Chapter 2 and Chapter 3 explain.

Inferential Statistics

CONCEPT   The branch of statistics that analyzes sample data to reach conclusions about a population.

EXAMPLE   A survey that sampled 1,264 women found that 45% of those polled considered friends or family their most trusted shopping advisers, and only 7% considered advertising their most trusted shopping adviser. By using methods discussed in Section 6.4, you can use these statistics to draw conclusions about the population of all women.

INTERPRETATION   When you use inferential statistics, you start with a hypothesis and look to see whether the data are consistent with that hypothesis. This deeper level of analysis means that inferential statistical methods can be easily misapplied or misconstrued and that many inferential methods require a calculating or computing device. (Chapters 6 through 9 discuss some of the inferential methods that you will most commonly encounter.)

1.4 Sources of Data

You begin every statistical analysis by identifying the source of the data that you will use for **data collection**. Among the important sources of data are **published sources**, **experiments**, and **surveys**.

Published Sources

CONCEPT   Data available in print or in electronic form, including data found on Internet websites. Primary data sources are those published by the individual or group that collected the data. Secondary data sources are those compiled from primary sources.

EXAMPLE   Many U.S. federal agencies, including the Census Bureau, publish primary data sources that are available at the data.gov website. Industry-specific groups and business news organizations commonly publish online or in-print secondary source data compiled by business organizations and government agencies.

INTERPRETATION   You should always consider the possible bias of the publisher and whether the data contain all the necessary and relevant variables when using published sources. This is especially true of sources found through Internet search engines.

Experiments

CONCEPT   A study that examines the effect on a variable of varying the value(s) of another variable or variables while keeping all other things equal. A typical experiment contains both a treatment group and a control group. The treatment group consists of those individuals or things that receive the treatment(s) being studied. The control group consists of those individuals or things that do not receive the treatment(s) being studied.

EXAMPLE   Pharmaceutical companies use experiments to determine whether a new drug is effective. A group of patients who have many similar character-istics is divided into two subgroups. Members of one group, the treatment group, receive the new drug. Members of the other group, the control group, often receive a placebo, a substance that has no medical effect. After a time period, statistics that describe each group are compared.

INTERPRETATION   Proper experiments are either single-blind or double-blind. A study is a single-blind experiment if only the researcher conducting the study knows the identities of the members of the treatment and control groups. If neither the researcher nor study participants know who is in the treatment group and who is in the control group, the study is a double-blind experiment.

When conducting experiments that involve placebos, researchers also have to consider the placebo effect—that is, whether people in the control group will improve because they believe they are getting a real substance that is intended to produce a positive result. When a control group shows as much improvement as the treatment group, a researcher can conclude that the placebo effect is a signif-icant factor in the improvements of both groups.

Surveys

CONCEPT   A process that uses questionnaires or similar means to gather values for the responses from a set of participants.

EXAMPLES   The decennial U.S. census mail-in form, a poll of likely voters, a website instant poll or "question of the day."

INTERPRETATION   Surveys are either **informal**, open to anyone who wants to participate; **targeted**, directed toward a specific group of individuals; or include people chosen at random. The type of survey affects how the data collected can be used and interpreted.

1.5 Sampling Concepts

The Section 1.2 definition of statistic notes that calculating statistics for a sample is the most common activity because collecting population data is usually impractical. Because samples are so commonly used, you need to learn the concepts that help identify all the members of a population and that describe how samples are formed.

Frame

CONCEPT The list of all items in the population from which the sample will be selected.

EXAMPLES Voter registration lists, municipal real estate records, customer or human resources databases, directories.

INTERPRETATION Frames influence the results of an analysis, and using different frames can lead to different conclusions. You should always be careful to make sure your frame completely represents a population; otherwise, any sample selected will be biased, and the results generated by analyses of that sample will be inaccurate.

Sampling

CONCEPT The process by which members of a *population* are selected for a *sample*.

EXAMPLES Choosing every fifth voter who leaves a polling place to interview, selecting playing cards randomly from a deck, polling every tenth visitor who views a certain website today.

INTERPRETATION Some sampling techniques, such as an "instant poll" found on a web page, are naturally suspect as such techniques do not depend on a well-defined frame. The sampling technique that uses a well-defined frame is *probability sampling*.

Probability Sampling

CONCEPT A sampling process that considers the chance of selection of each item. Probability sampling increases your chance that the sample will be representative of the population.

EXAMPLES The registered voters selected to participate in a recent survey concerning their intention to vote in the next election, the patients selected to fill out a patient-satisfaction questionnaire, 100 boxes of cereal selected from a factory's production line.

INTERPRETATION   You should use probability sampling whenever possible because *only* this type of sampling enables you to apply inferential statistical methods to the data you collect. In contrast, you should use nonprobability sampling, in which the chance of occurrence of each item being selected is not known, to obtain rough approximations of results at low cost or for small-scale, initial, or pilot studies that will later be followed up by a more rigorous analysis. Surveys and polls that invite the public to call in or answer questions on a web page are examples of nonprobability sampling.

Simple Random Sampling

CONCEPT   The probability sampling process in which every individual or item from a population has the same chance of selection as every other individual or item. Every possible sample of a certain size has the same chance of being selected as every other sample of that size.

EXAMPLES   Selecting a playing card from a shuffled deck or using a statistical device such as a table of random numbers.

INTERPRETATION   Simple random sampling forms the basis for other random sampling techniques. The word *random* in this phrase requires clarification. In this phrase, random means no repeating patterns—that is, in a given sequence, a given pattern is equally likely (or unlikely). It does not refer to the most commonly used meaning, "unexpected" or "unanticipated" (as in "random acts of kindness").

Other Probability Sampling Methods

Other, more complex, sampling methods are also used in survey sampling. In a stratified sample, the items in the frame are first subdivided into separate subpopulations, or strata, and a simple random sample is selected within each of the strata. In a cluster sample, the items in the frame are divided into several clusters so that each cluster is representative of the entire population. A random sampling of clusters is then taken, and all the items in each selected cluster or a sample from each cluster are then studied.

1.6   Sample Selection Methods

Sampling can be done either with or without replacement of the items being selected. Almost all survey sampling is done without replacement.

Sampling with Replacement

CONCEPT   A sampling method in which each selected item is returned to the frame from which it was selected so that it has the same probability of being selected again.

EXAMPLE   Selecting items from a fishbowl and returning each item to it after the selection is made.

Sampling Without Replacement

CONCEPT   A sampling method in which each selected item is not returned to the frame from which it was selected. Using this technique, an item can be selected no more than one time.

EXAMPLES   Selecting numbers in state lottery games, selecting cards from a deck of cards during games of chance such as blackjack or poker.

INTERPRETATION   Sampling without replacement means that an item can be selected no more than one time. You should choose sampling without replacement instead of sampling with replacement because statisticians generally consider the former to produce more desirable samples.

spreadsheet solution

Creating a New Worksheet and Entering Data

To create a new worksheet into which you can enter the data values of a variable for analysis, double-click the **Blank Workbook icon** in the New panel of the opening screen. If you have been using Excel and already have a worksheet open, select **File**, then **New** and, in the New panel, double-click the **Blank Workbook icon**.

To enter data into a specific cell of the new worksheet, move the cell pointer to that cell. You can move the pointer by either using the cursor keys, moving the mouse pointer, or completing the proper touch operation. As you type an entry, the entry appears in the formula bar area located over the top of the worksheet. You complete your entry by pressing **Tab** or **Enter** or by clicking the **checkmark button** in the formula bar.

To save your new file, select **File**, then **Save As** and, in the Save As dialog box, navigate to the folder where you want to save your file. Accept or revise the filename and then click **Save**. To later retrieve the file, select **File**, then **Open** and in the Open dialog box, navigate to the folder that contains the desired file, select the desired file from the list, and then click **Open**.

One-Minute Summary

Mastering basic vocabulary is the first step in learning statistics. Understanding the types of statistical methods, the sources of data used for data collection, sampling methods, and the types of variables used in statistical analysis are also important introductory concepts. Subsequent chapters focus on four important reasons for learning statistics:

- To present and describe information (Chapters 2 and 3)
- To reach conclusions about populations based only on sample results (Chapters 4 through 9)
- To develop reliable forecasts (Chapters 10 and 11)
- To use analytics to reach conclusions about large sets of data (Chapters 12 and 13)

Test Yourself

1. The portion of the population that is selected for analysis is called a:
    a. sample
    b. frame
    c. parameter
    d. statistic

2. A summary measure that is computed from only a sample of the population is called a:
    a. parameter
    b. population
    c. discrete variable
    d. statistic

3. The height of an individual is an example of a:
    a. discrete variable
    b. continuous variable
    c. categorical variable
    d. constant

4. The body style of an automobile (sedan, minivan, SUV, and so on) is an example of a:
    a. discrete variable
    b. continuous variable
    c. categorical variable
    d. constant

5. The number of credit cards in a person's wallet is an example of a:
   a.   discrete variable
   b.   continuous variable
   c.   categorical variable
   d.   constant

6. Statistical inference occurs when you:
   a.   compute descriptive statistics from a sample
   b.   take a complete census of a population
   c.   present a graph of data
   d.   take the results of a sample and reach conclusions about a population

7. The human resources director of a large corporation wants to develop a dental benefits package and decides to select 100 employees from a list of all 5,000 workers in order to study their preferences for the various components of a potential package. All the employees in the corporation constitute the _____.
   a.   sample
   b.   population
   c.   statistic
   d.   parameter

8. The human resources director of a large corporation wants to develop a dental benefits package and decides to select 100 employees from a list of all 5,000 workers in order to study their preferences for the various components of a potential package. The 100 employees who will participate in this study constitute the _____.
   a.   sample
   b.   population
   c.   statistic
   d.   parameter

9. Those methods that involve collecting, presenting, and computing characteristics of a set of data in order to properly describe the various features of the data are called:
   a.   statistical inference
   b.   the scientific method
   c.   sampling
   d.   descriptive statistics

10. Based on the results of a poll of 500 registered voters, the conclusion that the Democratic candidate for U.S. president will win the upcoming election is an example of:
   a.   inferential statistics
   b.   descriptive statistics

    c.   a parameter

    d.   a statistic

11. A numerical measure that is computed to describe a characteristic of an entire population is called a:

    a.   parameter

    b.   population

    c.   discrete variable

    d.   statistic

12. You wish to compare the value of the U.S. dollar to the English pound sterling. From a financial website, you obtain the values of the two currencies for the past 50 years. Which method of data collection were you using?

    a.   published sources

    b.   experimentation

    c.   surveying

13. Which of the following is a discrete variable?

    a.   The favorite flavor of ice cream of students at your local elementary school

    b.   The time it takes for a certain student to walk to your local elementary school

    c.   The distance between the home of a certain student and the local elementary school

    d.   The number of teachers employed at your local elementary school

14. Which of the following is a continuous variable?

    a.   The eye color of children eating at a fast-food chain

    b.   The number of employees of a branch of a fast-food chain

    c.   The temperature at which a hamburger is cooked at a branch of a fast-food chain

    d.   The number of hamburgers sold in a day at a branch of a fast-food chain

15. The number of cell phones in a household is an example of a:

    a.   categorical variable

    b.   discrete variable

    c.   continuous variable

    d.   statistic

Answer True or False:

16. The possible responses to the question "How long have you been living at your current residence?" are values from a continuous variable.

17. The possible responses to the question "How many times in the past seven days have you streamed a movie or TV show online?" are values from a discrete variable.

Fill in the Blank:

18. An insurance company evaluates many variables about a person before deciding on an appropriate rate for automobile insurance. The number of accidents a person has had in the past three years is an example of a _____ variable.

19. An insurance company evaluates many variables about a person before deciding on an appropriate rate for automobile insurance. The distance a person drives in a day is an example of a _____ variable.

20. An insurance company evaluates many variables about a person before deciding on an appropriate rate for automobile insurance. A person's marital status is an example of a _____ variable.

21. A numerical measure that is computed from only a sample of the population is called a _____.

22. The portion of the population that is selected for analysis is called the _____.

23. A college admission application includes many variables. The number of Advanced Placement courses the student has taken is an example of a _____ variable.

24. A college admission application includes many variables. The gender of the student is an example of a _____ variable.

25. A college admission application includes many variables. The distance from the student's home to the college is an example of a _____ variable.

Answers to Test Yourself

1. a	15. b
2. d	16. True
3. b	17. True
4. c	18. discrete
5. a	19. continuous
6. d	20. categorical
7. b	21. statistic
8. a	22. sample
9. d	23. discrete
10. a	24. categorical
11. a	25. continuous
12. a	
13. d	
14. c	

References

1. Berenson, M. L., D. M. Levine, K. A. Szabat, and D. F. Stephan. *Basic Business Statistics: Concepts and Applications*, 15th edition. Hoboken, NJ: Pearson Education, 2023.

2. Cochran, W. G. *Sampling Techniques*, 3rd edition. New York: John Wiley & Sons, 1977.

Presenting Data in Tables and Charts

Tables and charts are ways of summarizing categorical and numerical variables that can help you present information effectively. In this chapter, you will learn the appropriate types of tables and charts to use for each type of variable.

2.1 Presenting Categorical Variables

You present a categorical variable by first sorting values according to the categories of the variable. Then you place the count, amount, or percentage (part of the whole) of each category into a summary table or into one of several types of charts.

The Summary Table

CONCEPT   A two-column table in which category names are listed in the first column and the counts, amounts, or percentages of values are listed in a second column. Sometimes, additional columns present the same data in more than one way (for example, as counts and percentages).

EXAMPLE   A restaurant owner records the entrées ordered by guests during the Friday-to-Sunday weekend period. The data recorded can be presented using a summary table.

Entrée Ordered	Percentage
Beef	36
Chicken	26
Fish	28
Vegan	7
Other	3

INTERPRETATION   Summary tables enable you to see the big picture about a set of data. In this example, you can conclude that most customers will order beef, chicken, or fish. Very few will order either vegan or other entrées.

The Bar Chart

CONCEPT   A chart containing rectangles ("bars") in which the length of each bar represents the count, amount, or percentage of responses of one category.

EXAMPLE   The data of the summary table that the previous concept uses can be visualized using a percentage bar chart.

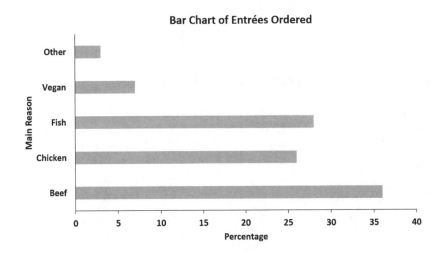

INTERPRETATION   A bar chart better presents the point that beef entrée is the single largest category of entrée ordered. For most people, scanning a bar chart is easier than scanning a column of numbers in which the numbers are unordered, as they are in the previous summary table.

The Pie Chart and the Doughnut Chart

CONCEPT

Pie: A circle chart in which wedge-shaped areas—pie slices—represent the count, amount, or percentage of each category, and the entire circle ("pie") represents the total.

Doughnut: A circle chart in which parts of the circumference represent the count, amount, or percentage of each category, and the entire circumference represents the total.

EXAMPLE   The following pie and doughnut charts visualize the summary table data that the two preceding concepts use.

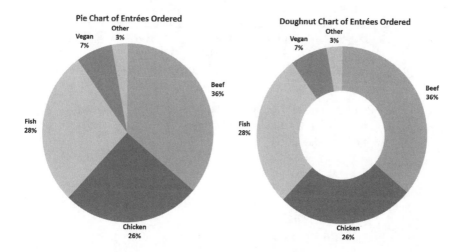

INTERPRETATION   A pie chart or a doughnut chart enables you to see how the various categories contribute to the whole. In the example charts, you can see that chicken and fish entrées make up about half of all entrées ordered and that beef is the entrée most ordered.

In recent years, doughnut charts have become preferred over pie charts. The *area* of pie "slices" can be misperceived, making the pie slice seem larger or smaller than the percentage of the whole that the slice represents. In contrast, doughnut charts focus attention on the lengths of each arc, which are easier to compare and accurately reflect the percentage of the whole.

Note that pie and doughnut charts do not enable you to as easily compare categories as a bar chart does. On the other hand, bar charts are less useful for understanding parts of a whole. The restaurant owner who recorded the entrée selections likely will want to compare categories and understand how each category contributes to the whole. Therefore, that person might use both a bar chart and a pie or doughnut chart to visualize the collected data.

spreadsheet solution

Bar, Pie, and Doughnut Charts

Chapter 2 Bar, **Chapter 2 Pie**, and **Chapter 2 Doughnut** present the preceding bar, pie, and doughnut charts, respectively. Experiment with each chart by entering your own values in column B of each worksheet that contains a chart.

Best Practices

Sort your summary table data by the values in the second column before you create a chart. This will enable you to create a chart that fosters comparisons. For a bar chart, arrange values from smallest to largest value if you want the longest bar to appear at the top of the chart; otherwise, sort the values from largest to smallest.

Reformat charts created by software to eliminate unwanted gridlines and legends or to change the text font and size of titles and axis labels.

How-Tos

Chart Tip CT1 (see Appendix D) explains how to sort data in a summary table.

Chart Tip CT2 lists common chart-reformatting commands.

Chart Tip CT3 lists the general steps for creating charts.

The Pareto Chart

CONCEPT   A special type of bar chart that presents the counts, amounts, or percentages of the categories, in descending order left to right, and also contains a superimposed plotted line that represents a running cumulative percentage.

EXAMPLE

Causes of Incomplete ATM Transactions

Cause	Frequency	Percentage
ATM malfunctions	32	4.42%
ATM out of cash	28	3.87%
Invalid amount requested	23	3.18%
Lack of funds in account	19	2.62%
Card unreadable	234	32.32%
Warped card jammed	365	50.41%
Wrong keystroke	23	3.18%
Total	724	100.00%

Source: Data extracted from A. Bhalla, "Don't Misuse the Pareto Principle," *Six Sigma Forum Magazine*, May 2009, pp. 15–18.

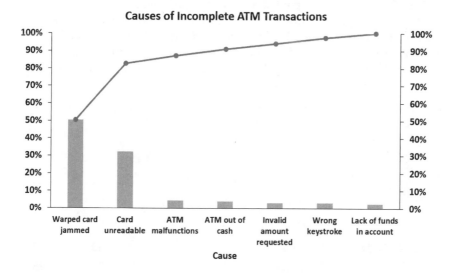

This Pareto chart uses the data of the table that immediately precedes it to highlight the causes of incomplete ATM transactions.

INTERPRETATION   When you have many categories, a Pareto chart enables you to focus on the most important categories by visually separating the *vital few* from the *trivial many* categories. For the incomplete ATM transactions data, the Pareto chart shows that two categories, warped card jammed and card unreadable, account for more than 80% of all defects and that those two categories combined with the ATM malfunctions and ATM out of cash categories account for more than 90% of all defects.

spreadsheet solution

Pareto Charts

Chapter 2 Pareto contains an example of a Pareto chart. Experiment with this chart by typing your own set of values—in descending order—in column B, rows 2 through 11. (Do not alter the entries in row 12 or columns C and D.)

How-To

Chart Tip CT4 (see Appendix D) summarizes how to create a Pareto chart.

Two-Way Table

CONCEPT   A table that presents the counts or percentages of responses for two categorical variables. In a two-way table, the categories of one of the variables form the rows of the table, while the categories of the second variable form the columns. The last row of a two-way table contains column totals, and the last column of such a table contains the row totals. Two-way tables are also known as cross-classification or cross-tabulation tables.

EXAMPLES   This two-way table tallies entrées ordered by guests during the Friday-to-Sunday weekend period by sex.

		Sex		
		Female	**Male**	**Total**
	Beef	64	80	144
	Chicken	53	51	104
Entrée Ordered	**Fish**	72	40	112
	Vegan	8	20	28
	Other	3	9	12
	Total	200	200	400

Two-way tables can be formatted to show grand total percentages or row or column percentages.

Grand Total Percentages Table

		Sex		
		Female	Male	Total
	Beef	16.00%	20.00%	36.00%
	Chicken	13.25%	12.75%	26.00%
Entrée Ordered	Fish	18.00%	10.00%	28.00%
	Vegan	2.00%	5.00%	7.00%
	Other	0.75%	2.25%	3.00%
	Total	50.00%	50.00%	100.00%

Row Percentages Table

		Sex		
		Female	Male	Total
	Beef	44.44%	55.56%	100.00%
	Chicken	50.96%	49.04%	100.00%
Entrée Ordered	Fish	64.29%	35.71%	100.00%
	Vegan	28.57%	71.43%	100.00%
	Other	25.00%	75.00%	100.00%
	Total	50.00%	50.00%	100.00%

Column Percentages Table

		Sex		
		Female	Male	Total
	Beef	32.00%	40.00%	36.00%
	Chicken	26.50%	25.50%	26.00%
Entrée Ordered	Fish	36.00%	20.00%	28.00%
	Vegan	4.00%	10.00%	7.00%
	Other	1.50%	4.50%	3.00%
	Total	100.00%	100.00%	100.00%

INTERPRETATION   The simplest two-way table contains a row variable that has two categories and a column variable that has two categories. This creates a table that has two rows and two columns in its inner part (see the table on the next page). Each inner cell represents the count or percentage of a pairing, or cross-classifying, of categories from each variable.

		Column Variable		
		First Column Category	Second Column Category	Total
Row Variable	First Row Category	Count or percentage for first row and first column categories	Count or percentage for first row and second column categories	Total for first row category
	Second Row Category	Count or percentage for second row and first column categories	Count or percentage for second row and second column categories	Total for second row category
	Total	Total for first column category	Total for second column category	Overall total

Two-way tables reveal the combination of values that occurs most often in data. In the example, the tables reveal that males are more likely to order beef than females and that females are more likely to order fish.

PivotTables create worksheet summary tables from sample data and provide a good way of creating two-way tables from sample data. Advanced Technique AT1 in Appendix E discusses how to create such tables.

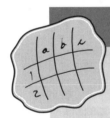

spreadsheet solution

Two-Way Tables

Chapter 2 Two-Way contains the counts of the download and call-to-action button variables as a simple two-way table.

Chapter 2 Two-Way PivotTable contains the counts of the entrée ordered and sex variables summarized in a two-way table that is an Excel PivotTable as well as PivotTables formatted to show grand total, row, and column percentage.

How-To

Advanced Technique ADV1 in Appendix E summarizes how to create a two-way table that is a PivotTable.

2.2 Presenting Numerical Variables

You present numerical variables by first establishing groups that represent separate ranges of values and then placing each value into the proper group. Then you create tables that summarize the groups by frequency (count) or percentage and use the table as the basis for creating charts such as a histogram, which this chapter explains.

The Frequency and Percentage Distribution

CONCEPT   A table of grouped numerical data that contains the names of each group in the first column, the counts (frequencies) of each group in the second column, and the percentages of each group in the third column. This table can also appear as a two-column table that shows either the frequencies or the percentages.

EXAMPLE   Consider the following data table, which presents the average ticket cost (in U.S. $) for each NBA team during a recent season.

NBA Ticket Cost

Team	Average Ticket Cost	Team	Average Ticket Cost
Atlanta	143	Miami	187
Boston	234	Milwaukee	153
Brooklyn	212	Minnesota	107
Charlotte	89	New Orleans	48
Chicago	251	New York	285
Cleveland	135	Oklahoma City	199
Dallas	124	Orlando	127
Denver	152	Philadelphia	197
Detroit	135	Phoenix	61
Golden State	463	Portland	119
Houston	177	Sacramento	198
Indiana	130	San Antonio	195
L.A. Clippers	137	Toronto	180
L.A. Lakers	444	Utah	78
Memphis	104	Washington	138

Source: Data extracted from "The Most Expensive NBA Teams to See Live," https://bit.ly/3rvSAah.

The following frequency and percentage distribution summarizes these data using 10 groupings from 0 to under 50 to 450 to under 500.

Average Ticket Cost	Frequency	Percentage
0 to under 50	1	3.33%
50 to under 100	3	10.00%
100 to under 150	11	36.67%
150 to under 200	9	30.00%
200 to under 250	2	6.67%
250 to under 300	2	6.67%
300 to under 350	0	0%
350 to under 400	0	0%
400 to under 450	1	3.33%
450 to under 500	1	3.33%
	30	100.00%

INTERPRETATION   Frequency and percentage distributions enable you to quickly determine differences among the many groups of values. In this example, you can quickly see that most of the average ticket costs are between $100 and $300 and that very few average ticket costs are either below $50 or above $200.

You need to be careful in forming distribution groups because the ranges of the groups affect how you perceive the data. For example, had you grouped the average ticket costs into only two groups, below $150 and $150 and above, you would not be able to see any pattern in the data.

Histogram

CONCEPT   A special bar chart for grouped numerical data in which the groups are represented as individual bars on the horizontal X axis and the frequencies or percentages for each group are plotted on the vertical Y axis. In a histogram, in contrast to a bar chart of categorical data, no gaps exist between adjacent bars.

EXAMPLE   The following histogram presents the average ticket cost data of the preceding example. The value below each bar (25, 75, 125, 175, 225, 275, 325, 375, 425, and 475) is the **midpoint**—the approximate middle value for the group the bar represents. As with the frequency and percentage distributions, you can quickly see that very few average ticket prices are above $275.

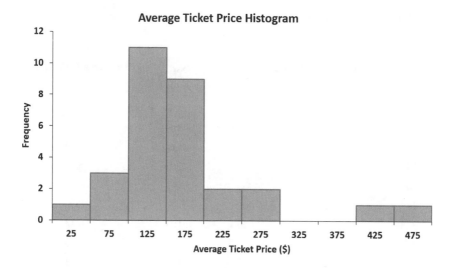

Average Ticket Price Histogram

INTERPRETATION A histogram reveals the overall shape of the frequencies in the groups. A histogram is considered symmetric if each side of the chart is an approximate mirror image of the other side. The histogram of this example has more values in the lower portion than in the upper portion, so it is considered to be non-symmetric, or *skewed*.

spreadsheet solution

Frequency Distributions and Histograms

Chapter 2 Histogram contains a frequency distribution and histogram for the average ticket cost (in U.S. $) for each NBA team during a recent season. Experiment with this chart by entering different values in column B, rows 3 through 12 of the Histogram worksheet.

How-Tos

Advanced Technique ADV2 in Appendix E and Chart Tip CT5 in Appendix D discuss how you can create frequency distributions and histograms.

The Time-Series Plot

CONCEPT A chart in which each point represents the value of a numerical variable at a specific time. By convention, the X axis (the horizontal axis) always represents units of time, and the Y axis (the vertical axis) always represents units of the variable.

EXAMPLE   Consider the following data table, which presents the number of domestic movie releases from 1990 to 2020.

Movie Releases

Year	Movies Released	Year	Movies Released
1990	224	2006	608
1991	244	2007	631
1992	234	2008	607
1993	258	2009	520
1994	254	2010	538
1995	279	2011	601
1996	310	2012	669
1997	303	2013	687
1998	336	2014	708
1999	384	2015	708
2000	371	2016	737
2001	355	2017	740
2002	480	2018	873
2003	507	2019	792
2004	551	2020	200
2005	547		

Source: Data extracted from "Domestic Yearly Box Office," https://www.boxofficemojo.com/year/.

The following time-series plot visualizes these data.

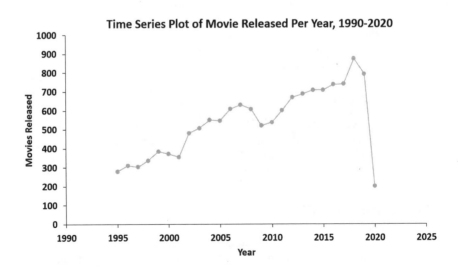

INTERPRETATION   Time-series plots can reveal patterns over time—patterns that you might not see when looking at a long list of numerical values. In this example, the plot reveals that, overall, there was a general increase in the number of movies released between 1990 and 2019. Before the steep drop in 2020 caused by the COVID-19 pandemic, the number of movies released in the preceding 30 years had increased fourfold.

The Scatter Plot

CONCEPT   A chart that plots the values of two numerical variables for each observation. In a scatter plot, the X axis (the horizontal axis) always represents units of one variable, and the Y axis (the vertical axis) always represents units of the second variable.

EXAMPLE   Consider the following data table, which presents the average ticket cost (in U.S. $) and the premium ticket cost (in U.S. $) for each NBA team during a recent season.

NBA Ticket Cost

Team	Average Ticket Cost	Premium Ticket Cost
Atlanta	143	267
Boston	234	448
Brooklyn	212	391
Charlotte	89	173
Chicago	251	493
Cleveland	135	268
Dallas	124	245
Denver	152	296
Detroit	135	266
Golden State	463	874
Houston	177	346
Indiana	130	252
L.A. Clippers	137	271
L.A. Lakers	444	857
Memphis	104	203
Miami	187	371
Milwaukee	153	301
Minnesota	107	204
New Orleans	48	89
New York	285	561
Oklahoma City	199	390
Orlando	127	249

Team	Average Ticket Cost	Premium Ticket Cost
Philadelphia	197	383
Phoenix	61	110
Portland	119	233
Sacramento	198	380
San Antonio	195	384
Toronto	180	338
Utah	78	142
Washington	138	271

The following scatter plot visualizes these data.

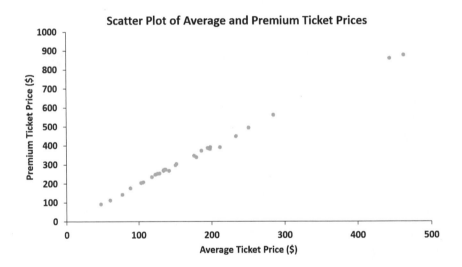

INTERPRETATION   A scatter plot helps reveal patterns in the relationship between two numerical variables. The scatter plot for these data reveals a strong positive linear (straight-line) relationship between the average ticket cost and the cost of a premium ticket. Based on this relationship, you can conclude that the average ticket cost is a useful predictor of the premium ticket cost. (Chapter 10 more fully discusses using one numerical variable to predict the value of another numerical variable.)

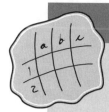

spreadsheet solution

Time-Series and Scatter Plots

Chapter 2 Time-Series contains the time-series plot for the domestic movie releases from 1990 to 2020. Experiment with this plot by entering different values in column B, rows 2 through 32.

Chapter 2 Scatter Plot contains the scatter plot for the NBA ticket cost data. Experiment with this scatter plot by entering different data values in columns B and C, rows 2 through 31.

How-Tos

Chart Tip CT6 (in Appendix D) discusses how you can create time-series plots.

Chart Tip CT7 (in Appendix D) discusses how you can create scatter plots.

2.3 **"Bad" Charts**

So-called "good" charts, such as the charts presented so far in this chapter, help visualize data in ways that aid understanding. However, in the modern world, you can easily find examples of "bad" charts that obscure or confuse the data. Such charts include elements or practices known to impede understanding or fail to apply properly the techniques that this chapter discusses.

CONCEPT   A "bad" chart fails to clearly present data in a useful and undistorted manner.

INTERPRETATION   Using pictorial symbols obscures the data and can create a false impression in the mind of the reader, especially if the pictorial symbols are representations of three-dimensional objects. In Example 1, the wine glasses fail to reflect that the 1992 data (2.25 million gallons) is a bit more than twice the 1.04 million gallons for 1989. In addition, the spaces between the wine glasses falsely suggest equal-sized time periods and obscure the trend in wine exports. (Hint: Plot the data as a time-series chart to discover the actual trend.)

EXAMPLE 1: Australian Wine Exports to the United States.

We're drinking more...
Australian wine exports to the U.S.
in millions of gallons

1.04   2.25   3.67   6.77

1989   1992   1995   1997

Example 2 combines the inaccuracy of using a picture (grape vine) with the error of having unlabeled and improperly scaled axes. A missing *X* axis prevents the reader from immediately seeing that the 1997–1998 value is misplaced. By the scale of the graph, that data point should be closer to the rest of the data. A missing *Y* axis prevents the reader from getting a better sense of the rate of change in land planted through the years. Other problems also exist. Can you spot at least one more? (Hint: Compare the 1949–1950 data to the 1969–1970 data.)

EXAMPLE 2: Amount of Land Planted with Grapes for the Wine Industry.

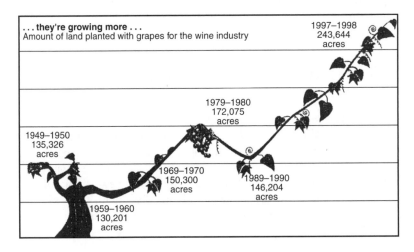

...they're growing more...
Amount of land planted with grapes for the wine industry

1997–1998
243,644
acres

1979–1980
172,075
acres

1949–1950
135,326
acres

1969–1970
150,300
acres

1989–1990
146,204
acres

1959–1960
130,201
acres

When producing your own charts, use these guidelines:

- Always choose the simplest chart that can present your data.
- Always supply a title.
- Always label every axis.
- Avoid unnecessary decorations or illustrations around the borders or in the background.
- Avoid the use of fancy pictorial symbols to represent data values.
- Avoid 3D versions of bar and pie charts.
- If the chart contains axes, always include a scale for each axis.
- When charting non-negative values, the scale on the vertical axis should begin at zero.

One-Minute Summary

To choose an appropriate table or chart type, begin by determining whether your data are categorical or numerical.

If your data are categorical:

- Determine whether you are presenting one or two variables.
- If one variable, use a summary table, bar chart, pie chart, or doughnut chart. If emphasizing the *vital few* from the *trivial many*, use a Pareto chart.
- If two variables, use a two-way table.

If your data are numerical:

- If charting one variable, use a frequency and percentage distribution with or without a histogram.
- If charting two variables, if the time order of the data is important, use a time-series plot; otherwise, use a scatter plot.

Test Yourself
Short Answers

1. Which of the following graphical presentations is not appropriate for categorical data?
    a.   Pareto chart
    b.   scatter plot

    c.   bar chart

    d.   pie chart

2. Which of the following graphical presentations is not appropriate for numerical data?

    a.   histogram

    b.   pie chart

    c.   time-series plot

    d.   scatter plot

3. A type of histogram in which the categories are plotted in the descending rank order of the magnitude of their frequencies is called a:

    a.   bar chart

    b.   pie chart

    c.   scatter plot

    d.   Pareto chart

4. Which of the following would best show that the total of all the categories sums to 100%?

    a.   pie chart

    b.   histogram

    c.   scatter plot

    d.   time-series plot

5. The basic principle behind the _____ is the capability to separate the vital few categories from the trivial many categories.

    a.   scatter plot

    b.   bar chart

    c.   Pareto chart

    d.   pie chart

6. When studying the simultaneous responses to two categorical variables, you should construct a:

    a.   histogram

    b.   pie chart

    c.   scatter plot

    d.   cross-classification table

7. In a cross-classification table, the number of rows and columns:

    a.   must always be the same

    b.   must always be two

    c.   must add to 100%

    d.   None of the above.

Answer True or False:

8. Histograms are used for numerical data, whereas bar charts are suitable for categorical data.

9. A website monitors customer complaints and organizes these complaints into six distinct categories. Over the past year, the company has received 534 complaints. One possible graphical method for representing these data is a Pareto chart.

10. A website monitors customer complaints and organizes these complaints into six distinct categories. Over the past year, the company has received 534 complaints. One possible graphical method for representing these data is a scatter plot.

11. A social media website collected information on the age of its customers. The youngest customer was 5, and the oldest was 96. To study the distribution of the age of its customers, the company should use a pie chart.

12. A social media website collected information on the age of its customers. The youngest customer was 5, and the oldest was 96. To study the distribution of the age of its customers, the company can use a histogram.

13. A website wants to collect information on the daily number of visitors. To study the daily number of visitors, it can use a pie chart.

14. A website wants to collect information on the daily number of visitors. To study the daily number of visitors, it can use a time-series plot.

15. A professor wants to study the relationship between the number of hours a student studied for an exam and the exam score achieved. The professor can use a time-series plot.

16. A professor wants to study the relationship between the number of hours a student studied for an exam and the exam score achieved. The professor can use a bar chart.

17. A professor wants to study the relationship between the number of hours a student studied for an exam and the exam score achieved. The professor can use a scatter plot.

18. If you wanted to compare the percentage of items that are in a particular category as compared to other categories, you should use a pie chart, not a bar chart.

Fill in the Blank:

19. To evaluate two categorical variables at the same time, a _____ should be developed.

20. A _____ is a vertical bar chart in which the rectangular bars are constructed at the boundaries of each class interval.

21. A _____ chart should be used when you are primarily concerned with the percentage of the total that is in each category.

22. A _____ chart should be used when you are primarily concerned with comparing the percentages in different categories.

23. A _____ should be used when you are studying a pattern between two numerical variables.

24. A _____ should be used to study the distribution of a numerical variable.

25. You have measured your pulse rate daily for 30 days. A _____ plot should be used to study the pulse rate for the 30 days.

26. You have collected data from your friends concerning their favorite soft drink. You should use a _____ chart to study the favorite soft drink of your friends.

27. You have collected data from your friends concerning the time it takes to get ready to leave their house in the morning. You should use a _____ to study this variable.

Answers to Test Yourself Short Answers

1. b
2. b
3. d
4. a
5. c
6. d
7. d
8. True
9. True
10. False
11. False
12. True
13. False
14. True
15. False
16. False
17. True
18. False
19. two-way table
20. histogram
21. pie chart
22. bar chart
23. scatter plot
24. histogram
25. time-series plot
26. bar chart, pie chart, or Pareto chart
27. histogram

Problems

1. A Pew Research Center survey studied the key issues for employed adults who have been working at home some or all of the time. The following three summary tables present the results of that survey.

Feeling Motivated to Do Their Work	Percentage
Very Difficult	7%
Somewhat Difficult	29%
Somewhat Easy	31%
Easy	34%

Doing Work Without Interruptions	Percentage
Very Difficult	8%
Somewhat Difficult	24%
Somewhat Easy	37%
Easy	31%

Having an Adequate Workspace	Percentage
Very Difficult	4%
Somewhat Difficult	19%
Somewhat Easy	31%
Easy	47%

For each table

a. Construct a bar chart and a pie or doughnut chart.

b. Which graphical method do you think best presents these data?

c. What conclusions can you reach concerning how employed adults who have been working at home some or all of the time feel about being motivated to do their work?

d. What conclusions can you reach concerning how employed adults who have been working at home some or all of the time feel about doing work without interruptions?

e. What conclusions can you reach concerning how employed adults who have been working at home some or all of the time feel about having an adequate workspace?

f. What differences in the responses among the three issues exist?

2. Market researchers for a telecommunications company have summarized data collected about the payment methods customers use in the following summary table.

Payment Method	Frequency
Bank transfer (automatic)	1,212
Credit card (automatic)	1,191
Electronic check	2,243
Mailed check	871
Total	5,517

a.   Using this table construct a bar chart and a pie or doughnut chart.

b.   Which graphical method do you think best presents these data?

c.   What conclusions can you reach about customer payment methods?

3. Medication errors are a serious problem in hospitals. The following summary table presents the root causes of pharmacy errors at a hospital during a recent time period.

Reason for Failure	Frequency
Additional instructions	16
Dose	23
Drug	14
Duplicate order entry	22
Frequency	47
Omission	21
Order not discontinued when received	12
Order not received	52
Patient	5
Route	4
Other	8

a.   Construct a Pareto chart for these data.

b.   Discuss the "vital few" and "trivial many" reasons for the root causes of pharmacy errors.

4. Students who attend a regional university located in a small town are known to favor the local independent pizza restaurant. A national chain of pizza restaurants looks to open a store in that town and conducts a survey of students who attend that university to determine pizza preferences. The following two-way table summarizes the survey variables store type and sex, based on the responses of a sample of 220 students.

		Sex	
		Female	Male
Store Type	Local	74	71
	National	19	56

a.   Construct a two-way table that displays grand total percentages.

b.   Construct a two-way table that displays row percentages.

c.   Construct a two-way table that displays column percentages.

d. What conclusions can you reach from the tables constructed in parts (a) through (c)?

e. Which table do you think is most useful in reaching the conclusions in your part (d) answer?

5. Churning, the loss of customers to a competitor, is a problem for all companies, especially telecommunications companies. Market researchers for a telecommunications company collect data from 5,517 customers of the company. Data collected for each customer includes whether the customer churned during the last month, the sex of the customer, whether the customer is a senior citizen, and whether the customer uses paperless billing. The following three summary tables summarize these survey variables.

		Churn	
		No	Yes
Sex	**Female**	1,858	883
	Male	1,903	873

		Churn	
		No	Yes
Senior Citizen	**No**	3,142	1,285
	Yes	619	471

		Churn	
		No	Yes
Paperless Billing	**No**	1,394	398
	Yes	2,367	1,358

For each table

a. Construct a two-way table that displays grand total percentages

b. Construct a two-way table that displays row percentages.

c. Construct a two-way table that displays column percentages.

d. What conclusions can you reach from the tables constructed in parts (a) through (c)?

e. Which table do you think is most useful in reaching the conclusions in your part (d) answer?

Domestic Beer

6. The file **Domestic Beer** contains the percentage alcohol, number of calories per 12 ounces, and number of carbohydrates (in grams) per 12 ounces for 157 of the best-selling domestic beers in the United States.

(Data extracted from "Find Out How Many Calories in Beer?" https://www.beer100.com/beer-calories.)

a.   Construct a frequency distribution and a percentage distribution for percentage alcohol, number of calories per 12 ounces, and number of carbohydrates per 12 ounces (in grams).

b.   Construct a histogram for percentage alcohol, number of calories per 12 ounces, and number of carbohydrates per 12 ounces (in grams).

c.   Construct three scatter plots: percentage alcohol versus calories, percentage alcohol versus carbohydrates, and calories versus carbohydrates.

d.   What conclusions can you reach about the percentage alcohol, number of calories per 12 ounces, and number of carbohydrates per 12 ounces (in grams)?

Super Bowl Ads

7. The **Super Bowl Ads** file contains the average ratings of 57 ads from the 2021 NFL Super Bowl broadcast. (Data extracted from T. Schad, "Rocket mortgage ads dominate Ad Meter," *USA Today*, February 9, 2021, p. 4B.)

a.   Construct a histogram based on these data.

b.   What conclusions can you reach concerning Super Bowl ad ratings?

8. The **Big Mac Starbucks** file contains the cost (in U.S. $) of a McDonald's Big Mac sandwich and a Starbucks tall latte in 11 world cities.

Big Mac Starbucks

City	Big Mac	Starbucks Tall Latte
Moscow	2.29	4.35
Johannesburg	2.53	2.18
Hong Kong	2.87	4.60
Bangkok	3.85	2.60
Dubai	4.08	4.29
Buenos Aires	4.22	2.14
London	4.32	3.58
New York	5.09	4.30
Paris	5.37	4.30
Toronto	4.38	3.15
Zurich	6.89	5.94

Source: Data extracted from "How Much a Big Mac Costs Around the World," *Business Insider*, https://businessinsider.com/mcdonalds-big-mac-price-around-the-world-2018-5, and "The Starbucks Index 2019," https://www.finder.com/starbucks.index.

a. Construct a scatter plot from these data.

b. What conclusions can you reach about the relationship between the cost of a McDonald's Big Mac and a Starbucks tall latte in these 11 world cities?

9. The **Potter Movies** file contains the first weekend gross (in $millions) and the total domestic gross (in $millions) for the eight movies in the Harry Potter film series.

Potter Movies

Title	First Weekend	Total Domestic
Sorcerer's Stone	90.295	317.871
Chamber of Secrets	88.357	262.233
Prisoner of Azkaban	93.687	249.758
Goblet of Fire	102.335	290.201
Order of the Phoenix	77.108	292.137
Half-Blood Prince	77.836	302.089
Deathly Hallows Part I	125.017	296.132
Deathly Hallows Part II	169.189	381.193

Source: Data extracted from "Box Office History for Harry Potter Movies," https://www.the-numbers.com/movies/franchise/Harry-Potter.

a. Construct a scatter plot from these data.

b. What conclusion can you reach about the relationship between the first weekend and total domestic grosses?

10. The **UHDTV Wholesale Sales** file contains the U.S. wholesale sales of Ultra HDTVs (in $millions) from 2013 to 2019.

UHDTV Wholesale Sales

Year	Wholesale Sales
2013	310
2014	2,238
2015	7,673
2016	12,932
2017	13,400
2018	14,300
2019	14,900

Source: Data extracted from "4K Ultra HD TVs wholesale sales revenue in the United States from 2013 to 2019," https://www.statista.com/statistics/643511/4k-ultra-hdtv-wholesale-sales-in-us/.

a. Construct a time-series plot of the U.S. Ultra HDTV wholesale sales from 2013 to 2019.

b. What pattern does the plot reveal?

c. If you were asked to predict U.S. Ultra HDTV wholesale sales for 2020, what would you predict?

11. The **MLB Salaries** file contains the average MLB baseball player salaries (in $millions) for the years 2003 through 2020.

MLB Salaries

Year	Average MLB Salary	Year	Average MLB Salary
2003	2.37	2012	3.21
2004	2.31	2013	3.39
2005	2.48	2014	3.69
2006	2.70	2015	3.84
2007	2.82	2016	4.38
2008	2.93	2017	4.45
2009	3.00	2018	4.41
2010	3.01	2019	4.80
2011	3.10	2020	4.43

Source: Data extracted from https://statista.com/statistics/23621/mean-salary-of-players-in-major-league-baseball (no longer available).

a. Construct a time-series plot of the average MLB baseball player salaries for the years 2003 through 2020.

b. What pattern does the plot reveal?

c. If you were asked to predict the average MLB baseball player salary for 2021, what would you predict?

Answers to Test Yourself Problems

1. b. If you are more interested in determining which category of feeling motivated to do their job response occurs most often, then the bar chart is preferred. If you are more interested in seeing the distribution of the entire set of categories, then either the pie chart or the doughnut chart is preferred.

c. Respondents are about equally likely to feel that it is easy, somewhat easy, or somewhat difficult to feel motivated to do their job.

d. Respondents are about equally likely to feel that it is somewhat easy or somewhat difficult to do work without interruption.

e. Respondents are most likely to feel that it is easy to have adequate workspace.

f. They feel that it is easier to have adequate workspace than to feel motivated to do work or to work without interruption.

2. b. If you are more interested in determining which category of payment method used occurs most often, then the bar chart is preferred. If you are more interested in seeing the distribution of the entire set of categories, either the pie chart or doughnut chart is preferred.

c. Respondents are most likely to pay by electronic check and least likely to pay by mailed check.

3. b. The most important categories of medication errors are orders not received and frequency followed by dose, duplicate order entry, and omission.

4. a. through c.

		Sex		
		Female	Male	Grand Total
Store Type	Local	33.64%	32.27%	65.91%
	National	8.64%	25.45%	34.09%
	Grand Total	42.28%	57.72%	100.00%

		Sex		
		Female	Male	Grand Total
Store Type	Local	51.03%	48.97%	100.00%
	National	25.33%	74.67%	100.00%
	Grand Total	42.27%	57.73%	100.00%

		Sex		
		Female	Male	Grand Total
Store Type	Local	79.57%	55.91%	65.91%
	National	20.43%	44.09%	34.09%
	Grand Total	100.00%	100.00%	100.00%

5. a. through c.
Sex and Churn

		Churn		
		No	Yes	Grand Total
Sex	Female	33.68%	16.01%	49.69%
	Male	34.49%	15.82%	50.31%
	Grand Total	68.17%	31.83%	100.00%

		Churn		
		No	**Yes**	**Grand Total**
Sex	**Female**	67.79%	32.21%	100.00%
	Male	68.55%	31.45%	100.00%
	Grand Total	68.17%	31.83%	100.00%

		Churn		
		No	**Yes**	**Grand Total**
Sex	**Female**	51.21%	50.59%	49.68%
	Male	48.79%	49.41%	50.32%
	Grand Total	100.00%	100.00%	100.00%

d. There is very little difference between males and females in churning.

e. Row percentages are more valuable because this table compares males and females.

Senior Citizen and Churn

		Churn		
		No	**Yes**	**Grand Total**
Senior Citizen	**No**	56.95%	23.29%	79.24%
	Yes	11.22%	8.54%	19.76%
	Grand Total	68.17%	31.83%	100.00%

		Churn		
		No	**Yes**	**Grand Total**
Senior Citizen	**No**	70.97%	29.03%	100.00%
	Yes	56.79%	43.21%	100.00%
	Grand Total	68.17%	31.83%	100.00%

		Churn		
		No	**Yes**	**Grand Total**
Senior Citizen	**No**	83.54%	73.17%	80.24%
	Yes	16.46%	26.83%	19.76%
	Grand Total	100.00%	100.00%	100.00%

d. Senior citizens are much less likely to churn.

e. Row percentages are more valuable because this table compares senior citizens and non-senior citizens.

Paperless Billing and Churn

		Churn		
		No	Yes	Grand Total
Paperless Billing	**No**	25.27%	7.21%	32.48%
	Yes	42.90%	24.61%	67.51%
	Grand Total	68.17%	31.62%	100.00%

		Churn		
		No	Yes	Grand Total
Paperless Billing	**No**	77.79%	22.21%	100.00%
	Yes	63.54%	36.46%	100.00%
	Grand Total	68.17%	31.83%	100.00%

		Churn		
		No	Yes	Grand Total
Paperless Billing	**No**	37.06%	22.67%	32.48%
	Yes	62.94%	77.33%	67.52%
	Grand Total	100.00%	100.00%	100.00%

d. Those who use paperless billing are more likely to churn than those who do not use paperless billing.

e. Row percentages are more valuable because this table best helps to compare those with and without paperless billing.

6. c. The alcohol percentage is concentrated between 4% and 6%, with more between 4% and 5%. The calories are concentrated between 140 and 160. The carbohydrates are concentrated between 12 and 15. There are outliers in the percentage of alcohol in both tails. The outlier in the lower tail is due to the nonalcoholic beer O'Doul's. The outlier in the upper tail is around 11.5%. A few beers have high calorie counts near 330 and carbohydrates as high as 32. A strong positive relationship exists between percentage of alcohol and calories and between calories and carbohydrates, and there is a moderately positive relationship between percentage alcohol and carbohydrates.

7. b.   The ad ratings are fairly symmetrical, with many of the ad scores between 5 and 6. Very few ratings are below 4.5 or above 7.

8. b.   There is a weak relationship between the cost of a McDonald's Big Mac and the cost of a Starbucks tall latte in various cities.

9. b.   There is a moderately positive relationship between the U.S. gross and the first weekend gross for Harry Potter movies.

10. b.   Ultra HDTV sales rose dramatically from 2013 to 2016 but leveled off after that.

    c.   Somewhere between 15 and 16 million.

11. b.   There has been a very strong linear increase in the salaries.

    c.   Because there was a decrease in 2020, the prediction is that the average salary in 2021 will be less than $5 million.

References

1. Beninger, J. M., and D. L. Robyn. 1978. "Quantitative Graphics in Statistics," *The American Statistician*, 32: 1–11.

2. Berenson, M. L., D. M. Levine, K. A. Szabat, and D. F. Stephan. *Basic Business Statistics: Concepts and Applications*, 15th edition. Hoboken, NJ: Pearson Education, 2023.

3. Levine, D., D. Stephan, and K. Szabat. *Statistics for Managers Using Microsoft Excel*, 9th edition. Boston: Pearson Education, 2021.

4. Tufte, E. R. *The Visual Display of Quantitative Information*, 2nd edition. Cheshire, CT: Graphics Press, 2002.

5. Tufte, E. R. *Visual Explanations*. Cheshire, CT: Graphics Press, 1997.

Descriptive Statistics

In addition to using the tables and charts that Chapter 2 discusses, you can also summarize and describe numerical variables by using descriptive measures that identify the properties of central tendency, variation, and shape.

3.1   Measures of Central Tendency

The data values for most numerical variables tend to group around a specific value. **Measures of central tendency** help describe to what extent this pattern holds for a specific numerical variable. This section discusses three commonly used measures: the arithmetic mean (also known as the mean or average), the median, and the mode. You calculate these measures as either sample statistics or population parameters.

The Mean

CONCEPT   A number equal to the sum of the data values, divided by the number of data values that were summed.

EXAMPLES   Many common sports statistics such as baseball batting averages and basketball points per game, mean score on a college entrance exam, mean age of the members of a social media website, mean waiting times at a bank.

INTERPRETATION   The mean represents one way of finding the most typical value in a set of data values. As the only measure of central tendency that uses all the data values in a sample or population, the mean has one great weakness: Individual extreme values can distort the most typical value, as WORKED-OUT PROBLEM 2 illustrates.

WORKED-OUT PROBLEM 1   Although many people sometimes find themselves running late as they get ready to go to work, few measure the actual time it takes to get ready in the morning. Suppose you want to determine the typical time that elapses between your alarm clock's programmed wake-up time and the time you leave your home for work. You decide to measure actual times (in minutes) for 10 consecutive working days and record the following times:

Times

Day	1	2	3	4	5	6	7	8	9	10
Time	39	29	43	52	39	44	40	31	44	35

To calculate the mean time, first sum the data values: 39 + 29 + 43 + 52 + 39 + 44 + 40 + 31 + 44 + 35, which equals 396. Then, take this sum of 396 and divide by 10, the number of data values. The result, 39.6 minutes, is the mean time to get ready.

WORKED-OUT PROBLEM 2   Consider the same problem but imagine that on day 4, an exceptional occurrence such as an unexpected phone call caused you to take 102 (and not 52) minutes to get ready in the morning. That would make the sum of all times 446 minutes, and the mean (446 divided by 10) would be 44.6 minutes.

This illustrates how one extreme value can dramatically change the mean. Instead of being a number at or near the middle of the 10 get-ready times, the new mean of 44.6 minutes is greater than 9 of the 10 get-ready times. In this case, the mean fails as a measure of a typical value, or "central tendency."

The Median

CONCEPT   The middle value when a set of data values has been ordered from lowest to highest value. When the number of data values is even, no natural middle value exists, and you calculate the mean of the two middle values to determine the median as the Interpretation that follows explains.

EXAMPLES   Economic statistics such as median household income for a region; marketing statistics such as the median age for purchasers of a consumer product; in education, the established middle point for many standardized tests.

equation
blackboard
(optional)

interested
in
math?

The WORKED-OUT PROBLEMS calculate the mean of a sample of get-ready times. You need three symbols to write the equation for calculating the mean:

- An uppercase italic X with a horizontal line above it, $\bar{X}$, pronounced as "X bar," that represents the number that is the mean of a sample.

- A subscripted uppercase italic X (for example, X_1) that represents one of the data values being summed. Because the problem contains 10 data values, there are 10 X values, the first labeled X_1 and the last labeled X_{10}.

- A lowercase italic n that represents the number of data values that were summed in this sample, a concept known as the **sample size**. You pronounce n as "sample size" to avoid confusion with the symbol N, which represents (and is pronounced as) the population size.

Using these symbols, the equation for calculating the mean can be expressed as:

$$\bar{X} = \frac{X_1 + X_2 + X_3 + X_4 + X_5 + X_6 + X_7 + X_8 + X_9 + X_{10}}{n}$$

By using an ellipsis (…), you can abbreviate the equation as:

$$\bar{X} = \frac{X_1 + X_2 + \cdots + X_{10}}{n}$$

Using the insight that the value of the last subscript will always be equal to the value of n, you can generalize the formula as:

$$\bar{X} = \frac{X_1 + X_2 + \cdots + X_n}{n}$$

By using the uppercase Greek letter sigma, Σ, a standard symbol that is used in mathematics to represent the summing of values, you can further simplify the formula as:

$$\bar{X} = \frac{\Sigma X}{n}$$

or more explicitly as:

$$\bar{X} = \frac{\displaystyle\sum_{i=1}^{n} X_i}{n}$$

in which i represents a placeholder for a subscript and the $i = 1$ and n below and above the sigma represent the numerical range of the subscripts to be used in the calculation.

INTERPRETATION   The median splits the set of ordered data values into two parts that have equal numbers of values. The median is a good alternative to the mean when extreme data values occur because, unlike with the mean, extreme values do not affect the median.

When you summarize an odd number of values, you calculate the median as the middle value of the ordered lowest-to-highest list of all data values. For example, if you had five ordered values, the median would be the third ordered value.

When you summarize an even number of data values, you calculate the median by taking the mean of the two values closest to the middle of the ordered lowest-to-highest list of all data values. For example, if you had six ordered values, you would calculate the mean of the third and fourth values. If you had ten ordered values, you would calculate the mean of the fifth and sixth values. (Some people use the term *ranks* to refer to ordered values, so the statement "If you had ten ranked values, you would calculate the mean of the fifth and sixth ranks" is equivalent to the previous sentence.)

important point

When a numerical variable has a very large number of data values, you cannot easily identify by visual inspection the middle value (or the middle two values, if the number of data values is even). When faced with a very large number of data values, add 1 to the number of data values and divide that sum by 2 to identify the position, or *rank*, of the median value in the ordered list of data values for the variable.

For example, if you had 127 ordered data values, you would divide 128 by 2 to get 64 and identify the median as the 64th ranked value. If you had 70 ordered values, you would divide 71 by 2 to get 35.5 and determine that the median is the mean of the 35th and 36th ranked values.

WORKED-OUT PROBLEM 3   You need to determine the median age of a group of employees whose individual ages are 47, 23, 34, 22, and 27. You calculate the median by first ranking the ages from lowest to highest: 22, 23, 27, 34, and 47. Because you have five values, the middle is the third-ranked value, 27, making the median 27. This means that half the workers are 27 years old or younger and half the workers are 27 years old or older.

WORKED-OUT PROBLEM 4   You need to determine the median for the original set of ten get-ready times from WORKED-OUT PROBLEM 1. Ordering these values from lowest to highest, you have

Time	29	31	35	39	39	40	43	44	44	52
Ordered Position	1st	2nd	3rd	4th	5th	6th	7th	8th	9th	10th

Because an even number of data values exists, you calculate the mean of the two values closest to the middle—that is, the fifth and sixth ranked values, 39 and 40. The mean of 39 and 40 is 39.5, making the median 39.5 minutes for the set of ten times to get ready.

Using the *n* symbol that the previous Equation Blackboard defines, you can define the median as:

$$\text{Median} = \frac{n+1}{2}\text{th ranked value}$$

The Mode

CONCEPT   The value (or values) in a set of data values that appears most frequently.

EXAMPLES   The most common score on an exam, the most common number of items purchased in one transaction at a store, the commuting time that occurs most often.

INTERPRETATION   Some sets of data values have no mode because all the unique values appear the same number of times. Other sets of data values can have more than one mode, such as the get-ready times that several earlier worked-out problems use. This set contains two modes, 39 minutes and 44 minutes, because each of these values appears twice, and all other values appear once.

As with the median, extreme values do not affect the mode. However, unlike with the median, the mode can vary much more from sample to sample than can the median or the mean.

3.2 Measures of Position

A measure of position describes the relative position of a specific value to the other values of a numerical variable. Commonly used measures of position are the rank of a value and the quartiles.

Rank

CONCEPT   The number that corresponds to the position of a data value in an ordered list of values.

EXAMPLE   In the ordered list of values 12, 23, 24, and 49, the data value 12 has the rank of 1, 23 has the rank of 2, and 40 has the rank of 4.

INTERPRETATION   Certain statistics and statistical procedures, including quartiles, use the rank of a data value—and not the actual data value—in calculations. In everyday situations, people often refer to positions in ordered data values by using *ordinal* position, an equivalent concept. For example, in referring to the ordered list 12, 23, 24, and 49, most would say that 12 is the *first* value, 23 is the *second* value, and 49 is the *fourth* value.

Quartiles

CONCEPT   The three values that split a set of ranked data values into four equal parts, or quartiles. The **first quartile, Q_1,** is the value such that 25.0% of the ranked data values are smaller and 75.0% are larger. The **second quartile, Q_2,** splits the ranked values into two equal parts. (The second quartile is another name for the median, which Section 3.1 defines.) The **third quartile, Q_3,** is the value such that 75.0% of the ranked values are smaller and 25.0% are larger.

EXAMPLE   Standardized tests that report results in terms of quartiles.

INTERPRETATION   Quartiles help bring context to a particular value that is part of a large set of values. For example, learning that you scored 580 (out of 800) on a standardized test would not be as informative as learning that you scored in the third quartile? that is, in the top 25% of all scores.

To determine the ranked value that defines the first quartile, add 1 to the number of data values and divide that sum by 4. For example, for 11 values, divide 12 by 4 to get 3 to determine that the third ranked value is the first quartile. To determine the ranked value that defines the third quartile, add 1 to the number of data values, divide that sum by 4, and multiply the quotient by 3. For the example of 11 values, the ninth ranked value is the third quartile (12 divided by 4 is 3, and 3 times 3 is 9). To determine the ranked value that defines the second quartile, use the page 48 instructions for calculating the median.

When the result of a quartile rank calculation is not an integer:

1. Select the ranked value whose rank is immediately below the calculated rank and select the ranked value whose rank is immediately above the calculated rank. For example, if the result of a quartile rank calculation is 3.75, select the third and fourth ranked values.

2. If the two ranked values selected are the same number, then the quartile is that number. If the two values are different numbers, continue with steps 3 through 5.

3. Multiply the larger ranked value by the decimal fraction of the calculated rank. (The decimal fraction will be 0.25, 0.50, or 0.75.)

4. Multiply the smaller ranked value by 1 minus the decimal fraction of the calculated rank.

5. Add the two products to determine the quartile value.

For example, if you had 10 values, the calculated rank for the first quartile would be 2.75 (10 + 1 is 11, 11/4 is 2.75). Because 2.75 is not an integer, you would select the second and third ranked values. If these two values were the same, then the first quartile would be the shared value. Otherwise, according to steps 3 and 4, you multiply the third ranked value by 0.75 and multiply the second ranked value by 0.25 (1 – 0.75 is 0.25). Then, according to step 5, you would add these products together to get the first quartile.

WORKED-OUT PROBLEM 5   You are asked to determine the first quartile for the following ranked get-ready times.

Time	29	31	35	39	39	40	43	44	44	52
Rank	1	2	3	4	5	6	7	8	9	10

You first calculate the rank for the first quartile as 2.75 (10 + 1 is 11, 11/4 is 2.75). Because 2.75 is not an integer, you select the second and third ranked values. According to steps 3 and 4, you multiply the third ranked value, 35, by 0.75 to get 26.25 and multiply the second ranked value, 31, by 0.25 to get 7.75. According to step 5, you calculate the first quartile as 34 (26.25 + 7.75 is 34). This means that 25% of the get-ready times are 34 minutes or less and that the other 75% are 34 minutes or more.

WORKED-OUT PROBLEM 6   You are asked to determine the third quartile for the ranked get-ready times. Calculate the rank for the third quartile as 8.25 (10 + 1 is 11, 11/4 is 2.75, 2.75 times 3 is 8.25). Because 8.25 is not an integer, you select the eighth and ninth ranked values. According to step 2, because both of these values are 44, the third quartile is 44. (No multiplication is necessary.)

equation
blackboard
(optional)

interested
in
math?

Using the equation for the median developed earlier,

$$\text{Median} = \frac{n+1}{2} \text{ranked value,}$$

you can express the **first quartile, Q_1,** as

$$Q_1 = \frac{n+1}{4} th \text{ ranked value}$$

and the **third quartile, Q_3,** as

$$Q_3 = \frac{3(n+1)}{4} th \text{ ranked value}$$

WORKED-OUT PROBLEM 7   You wish to further analyze the data in **Super Bowl Ads** that is first used in Problem 7 in Chapter 2. This file contains the average ratings of 57 ads from the 2021 NFL Super Bowl broadcast. For this analysis, you seek to compare the average ratings of advertisements shown during the first half of the Super Bowl with the average ratings of advertisements shown during the second half. Already, the data have been separated in two groups, and the average ratings of each group have been ordered lowest to highest as follows.

Super Bowl Ads

> **Ordered First Half Average Ratings**
>
> 4.5 4.7 5.0 5.0 5.2 5.2 5.3 5.3 5.3 5.3 5.3 5.4 5.5 5.8 5.8 5.9 6.0 6.1 6.2 6.2 6.3 6.4 6.4 6.5 6.7 6.7 6.7 7.4

> **Ordered Second Half Average Ratings**
>
> 4.0 4.3 4.7 4.8 4.8 4.9 4.9 5.2 5.2 5.3 5.3 5.3 5.4 5.5 5.6 5.6 5.7 5.7 5.8 5.8 5.9 5.9 6.0 6.0 6.1 6.3 6.5 6.7 7.3

Because you have two sets of ordered average ratings, you decide to compute selected measures of central tendency or position as the basis for comparison. You place the two sets of ordered values in separate columns and design a spreadsheet similar to the "Measures of Central Tendency and Position" Spreadsheet Solution that appears later in this section. Your finalized spreadsheet is the one in **Chapter 3 WorkedOut Problem 7**, a portion of which follows.

Super Bowl Ad Score Comparision		
Measures of Central Tendency		
	First Half	Second Half
Arithmetic Mean	5.7893	5.5345
Median	5.8	5.6
Mode	5.3	5.3
only the first mode is reported		
Measures of Position		
	First Half	Second Half
First Quartile	5.300	5.050
Third Quartile	6.375	5.950

Table 3.1 presents results of interest from this spreadsheet. All of the statistics calculated support the conclusion that the first half ad ratings are somewhat higher than the second half ad ratings.

TABLE 3.1
Results of interest from Super Bowl ad ratings study

Statistic	Interpretation
Mean ad score	First half ads have a slightly higher mean.
Median ad score	Median for first half ad ratings is slightly higher than median for second half ad ratings.

Statistic	Interpretation
Mode	The set of first half ad ratings has the same first mode as the set of second half ad ratings.
First quartile	The first quartile, Q_1, for the first half ad ratings is slightly higher than the first quartile for the second half ad ratings.
Third quartile	The third quartile, Q_3, for the first half ad ratings is higher than the third quartile for the second half ad ratings.

Percentile

CONCEPT   The percentage of ranked values that are lower than the result being reported.

EXAMPLE   By the definitions given earlier, the first quartile, Q_1, is the 25th percentile; the second quartile, Q_2, is the 50th percentile; and the third quartile, Q_3, is the 75th percentile. A score reported as being in the 99th percentile would be exceptional because that score is greater than 99% of all scores; that is, the score is in the top 1% of all scores.

INTERPRETATION   Percentiles are commonly used in cases in which one seeks to know how one data value compares to all other values. For example, the result of a standardized test often includes a percentile to show the relative placement of the test taker among all who took the test. Likewise, a school transcript often includes a percentile to show those reviewing the transcript the relative placement of the student among all students in the school.

spreadsheet solution

Measures of Central Tendency and Position

Chapter 3 Descriptive contains a spreadsheet (shown below) that calculates measures of central tendency and position for the sample of get-ready times. Experiment with the spreadsheet by entering your own data in column A.

Best Practices

Enter the data for the variable being summarized in its own column.

Use the **AVERAGE** (for the mean), **MEDIAN**, and **MODE** functions in worksheet formulas to calculate measures of central tendency.

Use the **QUARTILE.EXC** function to calculate the first and third quartiles.

How-Tos

Function Tip FT1 in Appendix D explains how to enter the formulas that **Chapter 3 Descriptive** uses.

Advanced Technique ATT2 in Appendix E describes the use of Analysis ToolPak as a second way to generate measures of central tendency.

Chapter 3 WorkedOut Problem 7 illustrates how to modify the Descriptive worksheet to compute measures for two groups.

	A	B	C	D	
1	Data		Descriptive Statistics		
2	29				
3	31		Measures of Central Tendency		
4	35		Arithmetic Mean	39.6	=AVERAGE(A:A)
5	39		Median	39.5	=MEDIAN(A:A)
6	39		Mode	39	=MODE(A:A)
7	40		*only the first mode is reported*		
8	43				
9	44		Measures of Position		
10	44		First Quartile	34.00	=QUARTILE.EXC(A:A,1)
11	52		Third Quartile	44.00	=QUARTILE.EXC(A:A,3)
12					
13			Measures of Variation		
14			Maximum	52	=MAX(A:A)
15			Minimum	29	=MIN(A:A)
16			Range	23	=D14 - D15
17			Variance	45.82	=VAR.S(A:A)
18			Standard Deviation	6.77	=STDEV.S(A:A)

3.3 Measures of Variation

Measures of **variation** show the amount of **dispersion**, or spread, in the data values of a numerical variable. Four frequently used measures of variation are the range, the variance, the standard deviation, and the Z score, all of which can be calculated as either sample statistics or population parameters.

The Range

CONCEPT   The difference between the largest and smallest values in a set of data values.

EXAMPLE   For the get-ready times data that several earlier worked-out problems use, the range is 23 minutes, which is the difference between 52 minutes, the largest value, and 29, the smallest value ($52 - 29$).

INTERPRETATION   The range is the number that represents the largest possible difference between any two values in a set of data values. The greater the range, the greater the variation in the data values.

WORKED-OUT PROBLEM 8   For the Super Bowl ad ratings study, the range for the first half ad ratings is 2.9, while the range for the second half ad ratings is 3.3. You can conclude that the second half ad ratings show more variation than the first half ad ratings.

The Variance and the Standard Deviation

CONCEPT   Two measures that describe how the set of data values for a variable are distributed around the mean of the variable. The standard deviation is the positive square root of the variance.

EXAMPLE   The variance among SAT scores for incoming freshmen at a college, the standard deviation of the time visitors spend on a website, the standard deviation of the annual return of a certain type of mutual funds.

INTERPRETATION   For almost all sets of data values, most values lie within an interval of plus and minus one standard deviation above and below the mean. Therefore, determining the mean and the standard deviation usually helps you define the range in which the majority of the data values occur.

interested in math?

To calculate the variance, you take the difference between each data value and the mean, square this difference, and then sum the squared differences. You then take this sum of squares (or *SS*) and divide it by 1 less than the number of data values for sample data or the number of data values for population data. The result is the variance.

Because calculating the variance includes squaring the difference between each value and the mean, a step that always produces a non-negative number, the variance can never be negative. And as the positive square root of such a non-negative number, the standard deviation can never be negative, either.

WORKED-OUT PROBLEM 9   You want to calculate the variance and standard deviation for the get-ready times that several earlier worked-out problems use. As first steps, in a table, you calculate the difference between each of the 10 individual times and the mean (39.6 minutes), square those differences, and sum the squares, as the table on the following page illustrates.

Day	Time	Difference: Time Minus the Mean	Square of the Difference
1	39	−0.6	0.36
2	29	−10.6	112.36
3	43	3.4	11.56

Day	Time	Difference: Time Minus the Mean	Square of the Difference
4	52	12.4	153.76
5	39	−0.6	0.36
6	44	4.4	19.36
7	40	0.4	0.16
8	31	−8.6	73.96
9	44	4.4	19.36
10	35	−4.6	21.16
			Sum of Squares: 412.40

Because this is a sample of get-ready times, the sum of squares, 412.40, is divided by 1 less than the number of data values, 9, to get 45.82, the sample variance. The square root of 45.82 (6.77, after rounding) is the sample standard deviation. You can then conclude that most get-ready times are between 32.83 (39.6 – 6.77) minutes and 46.37 (39.6 + 6.77) minutes, a statement that you can easily confirm for this *small* sample by visual examination.

WORKED-OUT PROBLEM 10   To continue the Super Bowl ad ratings study, you want to calculate the standard deviation for the first half and second half ad ratings. For first half ad ratings, the standard deviation is 0.7020, and you determine that the majority of first half ads will have a score between 5.0873 and 6.4913 (the mean 5.7893 ±0.7020). For second half ad ratings, the standard deviation is 0.7093, and you determine that the majority of those ads will have a score between 4.8252 and 6.2438 (the mean 5.5345 ±0.7093).

equation blackboard (optional)

interested in math?

Using symbols first introduced earlier in this chapter, you can express the sample variance and the sample standard deviation as:

$$\text{Sample variance} = S^2 = \frac{\sum (X_i - \overline{X})^2}{n-1}$$

$$\text{Sample standard deviation} = S = \sqrt{\frac{\sum (X_i - \overline{X})^2}{n-1}}$$

(continues on page 58)

spreadsheet solution

Measures of Variation

Starting in row 13, the **Chapter 3 Descriptive** spreadsheet, which the previous Spreadsheet Solution first mentions, also calculates measures of variation for the sample of get-ready times. Experiment with the spreadsheet by entering your own data in column A.

Best Practices

Enter the data for the variable being summarized in its own column.

Use the **VAR.S** and **STDEV.S** functions to calculate the variance and standard deviation for a sample. Use the **VAR.P** and **STDEV.P** functions to calculate the measures for a population.

Use the **MAX** and **MIN** functions to first identify the maximum and minimum values and then use a formula that takes the difference of those two values to calculate the range.

How-Tos

Function Tip FT1 in Appendix D explains how to enter the functions used in **Chapter 3 Descriptive**.

Advanced Technique ATT2 in Appendix E describes the use of Analysis ToolPak as a second way to generate measures of variation.

▲	A	B	C	D	
1	Data		Descriptive Statistics		
13			Measures of Variation		
14			Maximum	52	=MAX(A:A)
15			Minimum	29	=MIN(A:A)
16			Range	23	=D14 - D15
17			Variance	45.82	=VAR.S(A:A)
18			Standard Deviation	6.77	=STDEV.S(A:A))

To calculate the variance and standard deviation for population data, change the divisor from one less than the sample size (the number of data values in the sample) to the number of data values in the population, a value known as the **population size** and represented by an italicized uppercase N:

$$\text{Population variance} = \sigma^2 = \frac{\sum(X_i - \mu)^2}{N}$$

$$\text{Population standard deviation} = \sigma = \sqrt{\frac{\sum(X_i - \mu)^2}{N}}$$

In statistics, symbols in equations that represent population parameters are always Greek letters. Statistics uses the lowercase Greek letter sigma, σ, to represent the population standard deviation, replacing the uppercase italicized S for the sample standard deviation. Statistics also uses the lowercase Greek letter mu, μ, to represent the *population* mean, replacing the symbol $\overline{X}$ for the sample mean.

Standard (Z) Score

CONCEPT   The number that is the difference between a data value and the mean of the variable, divided by the standard deviation.

EXAMPLE   The Z score for a particular incoming freshman's SAT score, the Z score for the get-ready time on day 4.

INTERPRETATION   Z scores help you determine whether a data value is an extreme value, or *outlier*—that is, far from the mean. Generally, a data value's Z score that is less than –3 or greater than +3 indicates that the data value is an extreme value.

WORKED-OUT PROBLEM 11   You need to know whether any of the times from the set of ten get-ready times that several earlier worked-out problems use could be considered outliers. You calculate Z scores for each of those times using the previously calculated mean of 39.6 minutes and standard deviation of 6.77 minutes.

From these results, shown in the following table, you learn that the greatest positive Z score was 1.83 (for the day 4 value) and the greatest negative Z score was –1.27 (for the day 8 value). Because no Z score is less than –3 or greater than +3, you conclude that none of the get-ready times can be considered extreme.

Day	Time	Time Minus the Mean	Z Score
1	39	–0.6	–0.09
2	29	–10.6	–1.57
3	43	3.4	0.50
4	52	12.4	1.83
5	39	–0.6	–0.09
6	44	4.4	0.65
7	40	0.4	0.06
8	31	–8.6	–1.27
9	44	4.4	0.65
10	35	–4.6	–0.68

equation blackboard (optional)

Using symbols presented earlier in this chapter, you can express the Z score as:

$$Z \text{ score} = Z = \frac{X - \overline{X}}{S}$$

3.4 Shape of Distributions

Shape describes the pattern of the distribution of data values in a set of numerical data. There are three possibilities for the pattern: symmetric, left-skewed, or right-skewed. Shape is important as a set of data values that is too badly skewed can make certain statistical methods invalid, as a later chapter explains.

Symmetric Shape

CONCEPT   A set of data values in which the mean equals the median value, and the pattern of distribution of values to the left of the mean mirrors the pattern of distribution of values to the right.

EXAMPLE   Scores on a standardized exam, actual amount of soft drink in a one-liter bottle.

Left-Skewed Shape

CONCEPT   A set of data values in which the mean is less than the median value and the left tail of the distribution is longer than the right tail of the distribution. Also known as negative skew.

EXAMPLE   Scores on an exam in which most students score between 70 and 100, whereas a few students score between 10 and 69.

Right-Skewed Shape

CONCEPT   A set of data values in which the mean is greater than the median value and the right tail of the distribution is longer than the left tail of the distribution. Also known as positive skew.

EXAMPLE   Prices of homes in a particular community, annual family income.

INTERPRETATION   Right or positive skewness occurs when the set of data contains some extremely high data values. Left or negative skewness occurs when the set of data contains some extremely low values. The set of data values is symmetric when low and high values balance each other out.

When identifying shape, you should avoid the common pitfall of thinking that the side of the histogram in which most data values cluster closely together is the direction of the skew. For example, consider the three histograms that Figure 3.1 displays.

FIGURE 3.1

Histograms
of different
distributions
of data values

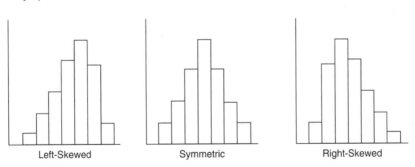

| Left-Skewed | Symmetric | Right-Skewed |

In the first histogram, the clustering of data values appears toward the right of the histogram, but the pattern is properly labeled left-skewed. To see the shape more clearly, statisticians create area-under-the-curve or distribution graphs, in which a plotted, curved line represents the tops of all the bars. Figure 3.2 shows these distribution graphs for the three Figure 3.1 histograms. If you remember that in such graphs the longer tail points to the skewness, you will never wrongly identify the direction of the skew.

FIGURE 3.2

Distribution
graphs
equivalent to
the Figure 3.1
histograms

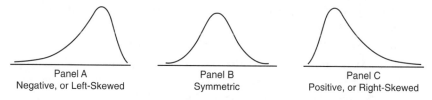

| Panel A | Panel B | Panel C |
| Negative, or Left-Skewed | Symmetric | Positive, or Right-Skewed |

In lieu of graphing a distribution, a skewness statistic can also be calculated. For this statistic, a value of 0 means a perfectly symmetric shape.

WORKED-OUT PROBLEM 12 You want to identify the shape of the NBA average ticket cost data. You examine the histogram of these data and determine that the distribution appears to be right-skewed because more low values than high values exist, and there are some very high values.

WORKED-OUT PROBLEM 13 The skewness for the get-ready times data is 0.086. Because this value is so close to 0, you conclude that the distribution of get-ready times around the mean is also approximately symmetric.

The Boxplot

CONCEPT A chart that plots the *five-number summary*, the five numbers that correspond to the smallest value, the first quartile Q_1, the median, the third quartile Q_3, and the largest value for a set of data. This chart is also known as a box-and-whisker plot.

INTERPRETATION The five-number summary concisely summarizes the shape of a set of data values. A boxplot visualizes the five numbers as vertical lines, interconnected to form a box shape from which a pair of lines extend (see Figure 3.3). From a boxplot, you can determine the degree of symmetry (or skewness) of a set of data based on the distances that separate the five numbers.

FIGURE 3.3

A generalized
boxplot

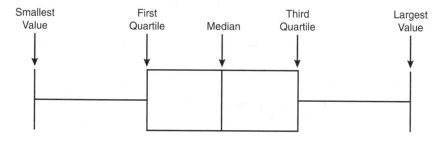

A boxplot for a set of data with a symmetric shape will contain the following relationships:

- The distance from the line that represents the smallest value to the line that represents the median equals the distance from the line that represents the median to the line that represents the largest value.

- The distance from the line that represents the smallest value to the line that represents the first quartile equals the distance from the line that represents the third quartile to the line that represents the largest value.
- The distance from the line that represents the first quartile to the line that represents the median equals the distance from the line that represents the median to the line that represents the third quartile.

A boxplot for a set of data with a right-skewed shape will contain the following relationships:

- The distance from the line that represents the median to the line that represents the largest value is greater than the distance from the line that represents the smallest value to the line that represents the median.
- The distance from the line that represents the third quartile to the line that represents the largest value is greater than the distance from the line that represents the smallest value to the line that represents the first quartile.
- The distance from the line that represents the first quartile to the line that represents the median is less than the distance from the line that represents the median to the line that represents the third quartile.

A boxplot for a set of data with a left-skewed shape will contain the following relationships:

- The distance from the line that represents the smallest value to the line that represents the median is greater than the distance from the line that represents the median to the line that represents the largest value.
- The distance from the line that represents the smallest value to the line that represents the first quartile is greater than the distance from the line that represents the third quartile to the line that represents the largest value.
- The distance from the line that represents the first quartile to the line that represents the median is greater than the distance from the line that represents the median to the line that represents the third quartile.

By these rules, the generalized boxplot in Figure 3.3 visualizes a set of data that is perfectly symmetric.

WORKED-OUT PROBLEM 14   You want to create a boxplot to summarize the set of times to get ready that WORKED-OUT PROBLEM 1 first uses. After calculating the five-number summary for this set of data, you create the Figure 3.4 boxplot. The boxplot seems to indicate an approximately symmetric shape to the set of get-ready times. The line that represents the median in the middle of the box is approximately equidistant between the ends of the box, and the lines from the median to the smallest and largest values do not greatly differ in length.

FIGURE 3.4

Boxplot for
the set of
get-ready
times

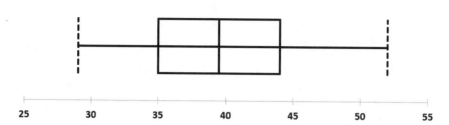

Boxplot for Get-Ready Times

WORKED-OUT PROBLEM 15   You seek to better understand the shape of the Super Bowl ad ratings that several earlier worked-out problems use. You create boxplots for the first half and second half ad scores that Figure 3.5 displays.

FIGURE 3.5

Boxplots for
the first half
and second
half Super
Bowl ad
ratings

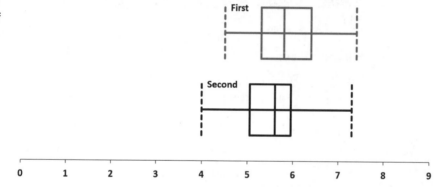

Boxplot of Super Bowl Ad Ratings by Half

The boxplot for the first half Super Bowl ad ratings shows that:

- The distance from the smallest value line to the median line is slightly less than the distance from the median line to the highest value line.
- The distance from the smallest value line to the first quartile line is less than the distance from the third quartile line to the highest value line.
- The distance from the first quartile line to the median line is slightly less than the distance from the median line to the third quartile line.

You conclude that the first half ad ratings are slightly right-skewed. The boxplot for the second half Super Bowl ad ratings shows that:

- The distance from the smallest value line to the median line is less than the distance from the median line to the highest value line.
- The distance from the smallest value line to the first quartile line is less than the distance from the third quartile line to the highest value line.

- The distance from the first quartile line to the median line is greater than the distance from the median line to the third quartile line.

Although there is some inconsistency, you conclude that the second half ad ratings are slightly right-skewed.

In comparing the first half and second half ad ratings, you conclude that the first half ad ratings are higher than the second half ad ratings because the minimum value, first quartile, median, third quartile, and maximum value are higher for the first half ad ratings.

spreadsheet solution

Measures of Shape

Chapter 3 Boxplot contains a boxplot for the get-ready times. Experiment with this chart by entering your own data in column A.

Best Practices

Use the **SKEW** function to calculate the skewness.

Use the **MIN**, **QUARTILE.EXC**, **MEDIAN**, and **MAX** functions to compute the five-number summary that serves as the source of the boxplot.

How-Tos

Function Tip FT1 in Appendix D explains how to enter the functions used in **Chapter 3 Descriptive**.

Advanced Technique ADV2 in Appendix E describes how to create results that include a measure of skewness.

Chapter 3 Boxplot uses a series of "tricked-up" line plots to create a boxplot using the calculated five-number summary in columns C and D and the formulas and data hidden behind the plot in columns F and G.

Note that the Excel box and whisker chart type creates a plot using different statistics and measures than the boxplot section of this chapter describes.

Chapter 3 WorkedOut Problem 15 explains the seven "tricked-up" line plots that form one boxplot.

Important Equations

Mean: $\bar{X} = \dfrac{\sum X_i}{n}$

Median: $\text{Median} = \dfrac{n+1}{2}th$ ranked value

First Quartile Q_1: $Q_1 = \dfrac{n+1}{4}th$ ranked value

Third Quartile Q_3: $Q_3 = \dfrac{3(n+1)}{4}th$ ranked value

Range: $\text{Range} = largest\ value - smallest\ value$

Sample Variance: $S^2 = \dfrac{\sum(X_i - \bar{X})^2}{n-1}$

Sample Standard Deviation: $S = \sqrt{\dfrac{\sum(X_i - \bar{X})^2}{n-1}}$

Population Variance: $\sigma^2 = \dfrac{\sum(X_i - \mu)^2}{N}$

Population Standard Deviation: $\sigma = \sqrt{\dfrac{\sum(X_i - \mu)^2}{N}}$

Z Scores: $Z = \dfrac{X - \bar{X}}{S}$

One-Minute Summary

The properties of central tendency, variation, and shape enable you to describe a set of data values for a numerical variable.

Numerical descriptive measures:

- Central tendency

    Mean

    Median

    Mode

- Variation

    Range
    Variance
    Standard deviation
    Z scores

- Shape

    Skew statistic
    Five-number summary
    Boxplot

Test Yourself
Short Answers

1. Which of the following statistics are measures of central tendency?

    a.   median
    b.   range
    c.   standard deviation
    d.   all of these
    e.   none of these

2. Which of the following statistics is not a measure of central tendency?

    a.   mean
    b.   median
    c.   mode
    d.   range

3. Which of the following statements about the median is not true?

    a.   It is less affected by extreme values than the mean.
    b.   It is a measure of central tendency.
    c.   It is equal to the range.
    d.   It is also known as the second quartile.

4. Which of the following statements about the mean is not true?

    a.   It is more affected by extreme values than is the median.
    b.   It is a measure of central tendency.
    c.   It is equal to the median in skewed distributions.
    d.   It is equal to the median in symmetric distributions.

5. Which of the following measures of variability is dependent on every value in a set of data?

   a.   range
   b.   standard deviation
   c.   each of these
   d.   neither of these

6. Which of the following statistics cannot be determined from a boxplot?

   a.   standard deviation
   b.   median
   c.   range
   d.   the first quartile

7. In a symmetric distribution:

   a.   the median equals the mean
   b.   the mean is less than the median
   c.   the mean is greater than the median
   d.   the median is less than the mode

8. The shape of a distribution is given by the:

   a.   mean
   b.   first quartile
   c.   skewness
   d.   variance

9. A five-number summary does not include the:

   a.   median
   b.   third quartile
   c.   mean
   d.   minimum (smallest) value

10. In a right-skewed distribution:

   a.   the median equals the mean
   b.   the mean is less than the median
   c.   the mean is greater than the median
   d.   the median equals the mode

Answer True or False:

11. In a boxplot, the box portion represents the data between the first and third quartile values.

12. The line drawn within the box of a boxplot represents the mean.

Fill in the Blanks:

13. The _____ is found as the middle value in a set of values placed in order from lowest to highest for an odd-sized sample of numerical data.

14. The standard deviation is a measure of _____.

15. If all the values in a data set are the same, the standard deviation will be _____.

16. A distribution that is negative-skewed is also called _____-skewed.

17. If each half of a distribution is a mirror image of the other half of the distribution, the distribution is called _____.

18. The median is a measure of _____.

19., 20., 21. The three characteristics that describe a set of numerical data are _____, _____, and _____.

For Questions 22 through 30, the number of days absent by a sample of nine students during a semester was as follows:

9   1   1   10   7   11   5   8   2

22. The mean is equal to _____.

23. The median is equal to _____.

24. The mode is equal to _____.

25. The first quartile is equal to _____.

26. The third quartile is equal to _____.

27. The range is equal to _____.

28. The variance is approximately equal to _____.

29. The standard deviation is approximately equal to _____.

30. The data are:
    a.   right-skewed
    b.   left-skewed
    c.   symmetric

31. In a left-skewed distribution:
    a.   the median equals the mean
    b.   the mean is less than the median
    c.   the mean is greater than the median
    d.   the median equals the mode

32. Which of the statements about the standard deviation is true?
    a.   It is a measure of variation around the mean.
    b.   It is the square of the variance.
    c.   It is a measure of variation around the median.
    d.   It is a measure of central tendency.

33. The smallest possible value of the standard deviation is _____.

Answers to Test Yourself Short Answers

1. a
2. d
3. c
4. c
5. b
6. a
7. a
8. c
9. c
10. c
11. True
12. False
13. median
14. variation
15. 0
16. left
17. symmetric

18. central tendency
19. central tendency
20. variation
21. shape
22. 6
23. 7
24. 1
25. 1.5
26. 9.5
27. 10
28. 15.25
29. 3.91
30. b
31. b
32. a
33. 0

Problems

MPG

1. The following data represent the overall miles per gallon (MPG) of 2021 sedans:

27 26 21 26 26 29 22 26 31 33 22 24 33 17
25 28 26 37 31 32 30 28 42 32 33 34 23

Source: Data extracted from "Sedans," https://www.consumerreports.org/cars/types/new/sedans/.

    a.   Compute the mean and median for these data.
    b.   Compute the first quartile and the third quartile for these data.
    c.   Compute the variance, standard deviation, and range for these data.
    d.   Construct a boxplot based on these data.
    e.   Are the data skewed? If so, how?
    f.   Based on the results of parts (a) through (e), what conclusions can you reach concerning the miles per gallon of midsized sedans?

NBA Ticket Cost

2. The following data table presents the average ticket cost for a single game for each NBA team during a recent season.

Team	Average Ticket Cost	Team	Average Ticket Cost
Atlanta	143	Miami	187
Boston	234	Milwaukee	153
Brooklyn	212	Minnesota	107
Charlotte	89	New Orleans	48
Chicago	251	New York	285
Cleveland	135	Oklahoma City	199
Dallas	124	Orlando	127
Denver	152	Philadelphia	197
Detroit	135	Phoenix	61
Golden State	463	Portland	119
Houston	177	Sacramento	198
Indiana	130	San Antonio	195
L.A. Clippers	137	Toronto	180
L.A. Lakers	444	Utah	78
Memphis	104	Washington	138

Source: Data extracted from "The Most Expensive NBA Teams to See Live," https://bit.ly/3zJeVUA.

a. Compute the mean and median for these data.
b. Compute the first quartile and the third quartile for these data.
c. Compute the variance, standard deviation, and range for these data.
d. Construct a boxplot based on these data.
e. Are the data skewed? If so, how?
f. Based on the results of (a) through (e), what conclusions can you reach about the average ticket price for NBA games?

Movies Released

3. The data table on page 71 contains the number of movie releases from 1990 to 2020.

a. Compute the mean and median for these data.
b. Compute the first quartile and the third quartile for these data.
c. Compute the variance, standard deviation, and range for these data.
d. Construct a boxplot based on these data.
e. Are the data skewed? If so, how?
f. Based on the results of (a) through (d), what conclusions can you reach about the yearly number of movie releases?

Year	Movies Released	Year	Movies Released
1990	224	1996	310
1991	244	1997	303
1992	234	1998	336
1993	258	1999	384
1994	254	2000	371
1995	279	2001	355
2002	480	2012	669
2003	507	2013	687
2004	551	2014	708
2005	547	2015	708
2006	608	2016	737
2007	631	2017	740
2008	607	2018	873
2009	520	2019	792
2010	538	2020	200
2011	601		

Source: Data extracted from "Domestic Yearly Box Office," https://www.boxofficemojo.com/year/.

Domestic Beer

4. The file **Domestic Beer** contains the percentage alcohol, number of calories per 12 ounces, and number of carbohydrates (in grams) per 12 ounces for 157 of the best-selling domestic beers in the United States. (Data extracted from "Find Out How Many Calories in Beer?" https://www.beer100.com/beer-calories.)

    a.   Compute the mean and median for these data.
    b.   Compute the first quartile and the third quartile for these data.
    c.   Compute the variance, standard deviation, and range for these data.
    d.   Construct a boxplot based on these data.
    e.   Are the data skewed? If so, how?
    f.   Based on the results of (a) through (e), what conclusions can you reach concerning the percentage alcohol, number of calories per 12 ounces, and number of carbohydrates (in grams) per 12 ounces?

Answers to Problems

1. a.   Mean = 28.2963, median = 28.
   b. $Q_1 = 25$, $Q_3 = 32$.
   c.   Variance = 28.9858, standard deviation = 5.3838, range = 25.

d. The mean for MPG is slightly more than the median, so the data are slightly right-skewed.

e. There is a longer tail on the right than on the left. The mean and median are about 28 MPG, with an extreme value of 42 MPG.

2. a. Mean = $173.40, median = $147.50.

b. Q_1 = $124, Q_3 = $198.

c. Variance = 8,752.6621, standard deviation = $93.5557, range = $415.

d. The mean is higher than the median. The difference between the largest value and Q_3 is $265, while the difference between Q_1 and the smallest value is $76, so the data are right-skewed.

e. There are several very high values that skew the mean. Half the average costs are above $147. The lowest average cost is $48.

3. a. Mean = 492.1290, median = 520.

b. Q_1 = 303, Q_3 = 669.

c. Variance = 38,906.4495, standard deviation = 197.3472, range = 673.

d. The mean is less than the median. There is a long upper tail due to an extreme value of 873, which makes the skewness inconsistent.

e. The mean is about 492, and half the years have more than 520 releases. The largest number of releases is 873, and the smallest number of releases is 200.

4. a. The means are percentage alcohol, 5.7695; calories, 155.6561; and carbohydrates, 12. The medians are percentage alcohol, 4.92; calories, 151; and carbohydrates, 8.65.

b. The first quartiles are percentage alcohol, 4.4; calories, 130.5; and carbohydrates, 8.65. The third quartiles are percentage alcohol, 5.65; calories, 170.5; and carbohydrates, 14.7.

c. The variances are percentage alcohol, 1.8344; calories, 1,917.4707; and carbohydrates, 24.8094. The standard deviations are percentage alcohol, 1.3544; calories, 43.7889; and carbohydrates, 4.9809. The ranges are percentage alcohol, 9.1; calories, 275; and carbohydrates, 30.2.

d. The percentage alcohol is right-skewed according to the boxplot, with a mean of 5.2695%. Half of the beers have less than 4.92% alcohol. The middle 50% of the beers have alcohol content spread over a range of 1.25%. The highest alcohol content is 11.5%, and the lowest is 2.4%.

e. The average scatter of percentage alcohol around the mean is 1.35044%. The number of calories is right-skewed according to the boxplot, with a mean of 155.6509. Half of the beers have calories below 151. The middle 50% of the beers have calories spread over a range of 40. The highest number of calories is 330, and the lowest is 55. The average scatter of calories around the mean is 43.7889. The number of carbohydrates is right-skewed according to the boxplot, with a mean of 12.0517, which is slightly higher than the median, at 12. Half of the beers have carbohydrates below 12. The middle 50% of the beers have carbohydrates spread over a range of 6.055. The highest number of carbohydrates is 32.1, and the lowest is 1.9. The average scatter of carbohydrates around the mean is 4.9809.

References

1. Berenson, M. L., D. M. Levine, K. A. Szabat, and D. F. Stephan. *Basic Business Statistics: Concepts and Applications*, 15th edition. Hoboken, NJ: Pearson Education, 2023.
2. Levine, D. M., D. F. Stephan, and K. A. Szabat, *Statistics for Managers Using Microsoft Excel*, 9th edition. Hoboken, NJ: Pearson Education, 2021.

Probability

Probability theory serves as one of the building blocks for inferential statistics. This chapter reviews the probability concepts necessary to comprehend the statistical methods discussed in later chapters. If you are already familiar with probability, you should skim this chapter to learn the vocabulary that this book uses to discuss probability concepts.

4.1   Events

Events underlie all discussions about probability. Before you can define probability in statistical terms, you need to understand the meaning of an event.

Event

CONCEPT   An outcome of an experiment or survey.

EXAMPLES   Rolling a die and turning up six dots, an individual who votes for the incumbent candidate in an election, someone using a social media website.

INTERPRETATION   Recall from Chapter 1 that performing experiments and conducting surveys are two important types of data sources. When discussing probability, many statisticians use the word *experiment* broadly to include surveys, so you can use the shorter definition "an outcome of an experiment" if you understand this broader usage of *experiment*. Likewise, as you read this chapter and encounter the word *experiment*, you should use the broader meaning.

Elementary Event

CONCEPT　An outcome that satisfies only one criterion.

EXAMPLES　A red card from a standard deck of cards, a voter who selected the Democratic candidate, a person who owns a 5G smartphone.

INTERPRETATION　Elementary events are distinguished from joint events, which meet two or more criteria.

Joint Event

CONCEPT　An outcome that satisfies two or more criteria.

EXAMPLES　A red ace from a standard deck of cards, a voter who voted for the Democratic candidate for president and the Democrat candidate for U.S. senator, a female who owns a 5G smartphone.

INTERPRETATION　Joint events are distinguished from elementary events, which meet only one criterion each.

4.2　More Definitions

Using the concept of event, you can define three more basic terms of probability.

Random Variable

CONCEPT　A variable whose numerical values represent the events of an experiment.

EXAMPLES　The number of text messages sent by a certain person in a 24-hour period, the scores of students on a standardized exam, the preferences of consumers for different brands of automobiles.

INTERPRETATION　You use the phrase **random variable** to refer to a variable that has no data values until an experimental trial is performed or a survey question is asked and answered. Random variables are either discrete, in which the possible numerical values are a set of integers (or coded values, in the case of categorical data), or continuous, in which any value is possible within a specific range.

Probability

CONCEPT   A number that represents the chance that a particular event will occur for a random variable.

EXAMPLES   Odds of winning a lottery, chance of rolling a seven when rolling two dice, likelihood of an incumbent winning reelection, chance that an individual will own a 5G smartphone.

INTERPRETATION   Probability determines the likelihood that a random variable will be assigned a specific value. Probability considers things that might occur in the future, and its forward-looking nature provides a bridge to inferential statistics.

Probabilities can be developed for an elementary event of a random variable or any group of joint events. For example, when rolling a standard six-sided die (see illustration), six possible elementary events correspond to the six faces of the die that contain either one, two, three, four, five, or six dots. "Rolling a die and turning up an even number of dots" is an example of an event formed from three elementary events (rolling a two, four, or six).

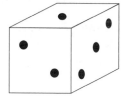

Probabilities are formally stated as decimal numbers in the range of 0 to 1. A probability of 0 indicates an event that never occurs. (Such an event is known as a **null event**.) A probability of 1 indicates a **certain event**—an event that must occur. For example, when you roll a die, getting seven dots is a null event because it can never happen, and getting six or fewer dots is a certain event because you will always end up with a face that has six or fewer dots. Probabilities can also be stated informally as the "percentage chance of (something)" or as quoted odds, such as a "50/50 chance."

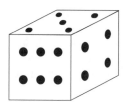

Collectively Exhaustive Events

CONCEPT   A set of events that includes all the possible events.

EXAMPLES   Heads and tails in the toss of a coin; the events "roll a value of 3 or less" and "roll a value of 4 or more" for a far six-sided die.

INTERPRETATION   When you have a set of collectively exhaustive events, one of them must occur. A coin must land on either heads or tails and a fair die must be "3 or less" or "4 or more." The sum of the individual probabilities associated with a set of collectively exhaustive events is always 1.

4.3 **Some Rules of Probability**

A set of rules governs the calculation of the probabilities of elementary and joint events.

RULE 1   The probability of an event must be between 0 and 1. The smallest possible probability value is 0. You cannot have a negative probability. The largest possible probability value is 1.0. You cannot have a probability greater than 1.0.

EXAMPLE   In the case of a die, the event of getting a face of seven has a probability of 0 because this event cannot occur. The event of getting a face with fewer than seven dots has a probability of 1.0 because it is certain that one of the elementary events of one, two, three, four, five, or six dots must occur.

RULE 2   The event that A does not occur is called **A complement**, or simply **not A**, and is given the symbol A'. If $P(A)$ represents the probability of event A occurring, then $1 - P(A)$ represents the probability of event A not occurring.

EXAMPLE   In the case of a die, the complement of getting the face that contains three dots is not getting the face that contains three dots. Because the probability of getting the face containing three dots is 1/6, the probability of not getting the face that contains three dots is $(1 - 1/6) = 5/6$ or 0.833.

RULE 3   If two events A and B are **mutually exclusive**, the probability of both events A and B occurring is 0. This means that the two events cannot occur at the same time.

EXAMPLE   On a single roll of a die, you cannot get a die that has a face with three dots and also have four dots because such elementary events are mutually exclusive. Either three dots can occur or four dots can occur, but not both.

RULE 4   If two events A and B are mutually exclusive, the probability of either event A or event B occurring is the sum of their separate probabilities.

EXAMPLE   The probability of rolling a die and getting either a face with three dots or a face with four dots is 1/3, or 0.333, because these two events are mutually exclusive. Therefore, the probability of 1/3 is the sum of the probability of rolling a three (1/6) and the probability of rolling a four (1/6), which is 1/3.

INTERPRETATION   You can extend this addition rule for mutually exclusive events to situations in which more than two events exist. In the case of rolling a die, the probability of turning up an even face (two, four, or six dots) is 0.50, the sum of 1/6 and 1/6 and 1/6 (3/6, or 0.50).

RULE 5   If events in a set are mutually exclusive and collectively exhaustive, the sum of their probabilities must add up to 1.0.

EXAMPLE   The events turning up a face with an even number of dots and turning up a face with an odd number of dots are mutually exclusive and collectively exhaustive. They are mutually exclusive because even and odd cannot occur simultaneously on a single roll of a die. They are also collectively exhaustive because either even or odd must occur on a particular roll. Therefore, for a single die, the probability of turning up a face with an even or odd face is the sum of the probability of turning up an even face plus the probability of turning up an odd face, or 1.0, as follows:

$$P \text{ (even or odd face)} = P \text{ (even face)} + P \text{ (odd face)}$$

$$= \frac{3}{6} + \frac{3}{6}$$

$$= \frac{6}{6} = 1$$

RULE 6   If two events A and B are not mutually exclusive, the probability of either event A or event B occurring is the sum of their separate probabilities minus the probability of their simultaneous occurrence (the joint probability).

EXAMPLE   For rolling a single die, turning up a face with an even number of dots is not mutually exclusive with turning up a face with fewer than five dots because both events include these (two) elementary events: turning up the face with two dots and turning up the face with four dots. To determine the probability of these two events, you add the probability of having a face with an even number of dots (3/6) to the probability of having a face with fewer than five dots (4/6) and then subtract the joint probability of simultaneously having a face with an even number of dots and having a face with fewer than five dots (2/6). You can express this as follows:

P (even face *or* face with fewer than five dots) =
P (even face) + P (face with fewer than five dots) −
P (even face *and* face with fewer than five dots)

$$= \frac{3}{6} + \frac{4}{6} - \frac{2}{6}$$

$$= \frac{5}{6}$$

$$= 0.833$$

INTERPRETATION   This rule requires that you subtract the joint probability because that probability has already been included twice (in the first event and in the second event). Because the joint probability has been "double counted," you must subtract it to compute the correct result.

RULE 7   If two events A and B are **independent**, the probability of both events A and B occurring is equal to the product of their individual probabilities. Two events are independent if the occurrence of one event in no way affects the probability of the second event.

EXAMPLE   When rolling a die, each roll of the die is an independent event because no roll can affect another (although gamblers who play dice games sometimes would like to think otherwise). Therefore, to determine the probability that two consecutive rolls both turn up the face with five dots, you multiply the probability of turning up that face on roll one (1/6) by the probably of turning up that face on roll two (also 1/6). You can express this as follows:

P (face with five dots on roll one and face with five dots on roll two) =

P (face with five dots on roll one) $\times P$ (face with five dots on roll two)

$$= \frac{1}{6} \times \frac{1}{6}$$
$$= \frac{1}{36} = 0.028$$

RULE 8   If two events A and B are not independent, the probability of both events A and B occurring is the product of the probability of event A multiplied by the probability of event B occurring, given that event A has occurred.

EXAMPLE   During the taping of a television game show, contestants are randomly selected from the audience watching the show. After a particular person has been chosen, he or she does not return to the audience and cannot be chosen again, therefore making this a case in which the two events are not independent.

If the audience consists of 30 women and 20 men (50 people), what is the probability that the first two contestants chosen are male? The probability that the first contestant is male is simply 20/50 or 0.40. However, the probability that the second contestant is male is not 20/50 because when the second selection is made, the eligible audience now has only 19 males and 49 people because the first male selected cannot be selected again. Therefore, the probability that the second selection is male is 19/49 or 0.388, rounded. This means that the probability that the first two contestants are male is 0.155, as follows:

P (male selection first and male selection second) =

P (male selection first) $\times P$ (male selection second)

$$= \frac{20}{50} \times \frac{19}{49}$$
$$= \frac{380}{2,450} = 0.155$$

4.4 Assigning Probabilities

Three different approaches exist for assigning probabilities to the events of a random variable: the classical approach, the empirical approach, and the subjective approach.

Classical Approach

CONCEPT   Assigning probabilities based on prior knowledge of the process involved.

EXAMPLE   Rolling a die and assigning the probability of turning up the face with three dots.

INTERPRETATION   Classical probability often assumes that all elementary events are equally likely to occur. When this is true, the probability that a particular event will occur is defined by the number of ways the event can occur divided by the total number of elementary events. For example, when you roll a die, the probability of getting the face with three dots is 1/6 because six elementary events are associated with rolling a die. Thus, you can expect that 1,000 out of 6,000 rolls of a die would turn up the face with three dots.

Empirical Approach

CONCEPT   Assigning probabilities based on frequencies obtained from empirically observed data.

EXAMPLE   Probabilities determined by polling or marketing surveys.

INTERPRETATION   The empirical approach does not use theoretical reasoning or assumed knowledge of a process to assign probabilities. Similar to the classical approach when all elementary events are equally likely, the empirical probability can be calculated by dividing the number of ways A can occur by the total number of elementary events. For example, if a poll of 500 registered voters reveals that 275 are likely to vote in the next election, you can assign the empirical probability of 0.55 (275 divided by 500).

Subjective Approach

CONCEPT   Assign probabilities based on expert opinions or other subjective methods such as "gut" feelings or hunches.

EXAMPLE   Commentators stating the odds that a political candidate will win an election or that a sports team will win a championship, a financial analyst stating the chance that a stock will increase in value by a certain amount in the next year.

INTERPRETATION   In this approach, you use your own intuition and knowledge or experience to judge the likeliest outcomes. You use the subjective approach when either the number of elementary events or actual data are not available for the calculation of relative frequencies. Because of the subjectivity, different individuals might assign different probabilities to the same event.

One-Minute Summary

Foundation concepts:

- Rules of probability
- Assigning probabilities

Test Yourself
Short Answers

1. If two events are collectively exhaustive, what is the probability that one or the other occurs?

    a.   0
    b.   0.50
    c.   1.00
    d.   The probability cannot be determined from the information given.

2. If two events are collectively exhaustive, what is the probability that both occur at the same time?

    a.   0
    b.   0.50
    c.   1.00
    d.   The probability cannot be determined from the information given.

3. If two events are mutually exclusive, what is the probability that both occur at the same time?

    a.   0
    b.   0.50
    c.   1.00
    d.   The probability cannot be determined from the information given.

4. If the outcome of event A is not affected by event B, then events A and B are said to be:

    a.   mutually exclusive
    b.   independent
    c.   collectively exhaustive
    d.   dependent

Use the following problem description when answering Questions 5 through 9:

A survey is taken among customers of a coffee shop to determine preference for regular or decaffeinated coffee. Of 200 respondents selected, 125 were male, and 75 were female. 120 preferred regular coffee, and 80 preferred decaffeinated coffee. Of the males, 85 preferred regular coffee.

5. The probability that a randomly selected individual is a male is:

   a.   125/200
   b.   75/200
   c.   120/200
   d.   200/200

6. The probability that a randomly selected individual prefers regular or decaffeinated coffee is:

   a.   0/200
   b.   125/200
   c.   75/200
   d.   200/200

7. Suppose that two individuals are randomly selected. The probability that both prefer regular coffee is:

   a.   (120/200)(120/200)
   b.   (120/200)
   c.   (120/200)(119/199)
   d.   (85/200)

8. The probability that a randomly selected individual prefers regular coffee is:

   a.   0/200
   b.   120/200
   c.   75/200
   d.   200/200

9. The probability that a randomly selected individual prefers regular coffee *or* is a male is:

   a.   0/200
   b.   125/200
   c.   160/200
   d.   200/200

10. The smallest possible value for a probability is _____.

11. The largest possible value for a probability is _____.

12. If two events are _____, they cannot occur at the same time.

13. If two events are _____, the probability that both events occur is the product of their individual probabilities.

14. In the _____ probability approach, probabilities are based on frequencies obtained from surveys.

15. In the _____ probability approach, probabilities can vary depending on the individual assigning them.

Answers to Test Yourself Short Answers

1. c

2. d

3. a

4. b

5. a

6. d

7. c

8. b

9. c

10. 0

11. 1

12. mutually exclusive

13. independent

14. empirical

15. subjective

Problems

1. Businesses use a method called *A/B testing* to test different web page designs to see if one design is more effective than another. For one company, designers were interested in the effect of modifying the call-to-action button on the home page. Every visitor to the company's home page was randomly shown either the original call-to-action button (the control) or the new variation. Designers measured success by the download rate: the number of people who downloaded the file divided by the number of people who saw that particular call-to-action button. Results of the experiment yielded the following:

		Call-to-Action Button		
		Original	New	Total
Download	Yes	351	451	802
	No	3,291	3,105	6,396
	Total	3,642	3,556	7,198

What is the probability that a respondent chosen at random

a. downloaded the file?

b. downloaded the file *and* used the new call-to-action button?

c. downloaded the file *or* used the new call-to-action button?

d. Suppose two respondents who used the original call-to-action button were selected. What is the probability that both downloaded the file?

2. Students who attend a regional university located in a small town are known to favor the local independent pizza restaurant. A national chain of pizza restaurants looks to open a store in that town and conducts a survey of students who attend that university to determine pizza preferences. The following summary table summarizes the survey variables store type and gender based on the responses of a sample of 220 students.

		Sex		
		Female	Male	Total
Store Type	Local	74	71	145
	National	19	56	75
	Total	93	127	220

What is the probability that a student chosen at random

a. prefers the local pizza restaurant?
b. is a female?
c. prefers the local pizza restaurant *and* is a female?
d. prefers the local pizza restaurant *or* is a female?
e. Given that the student selected is a female, what is the probability that she prefers the local pizza restaurant?
f. Given that the student selected is a male, what is the probability that he prefers the local pizza restaurant?
g. What do the results of parts (e) and (f) tell you about the difference between males and females concerning pizza preference?

3. Churning, the loss of customers to a competitor, is a problem for all companies, especially telecommunications companies. Market researchers for a telecommunications company collect data from 5,517 customers of the company. Data collected for each customer includes whether the customer churned during the last month and whether the customer uses paperless billing. The following summary table summarizes these survey variables.

		Churn		
		No	Yes	Total
Paperless Billing	No	1,394	398	1,792
	Yes	2,367	1,358	3,725
	Total	3,761	1,756	5,517

What is the probability that a customer selected at random

a. churned in the last month?
b. uses paperless billing?
c. uses paperless billing and churned in the last month?

d. uses paperless billing or churned in the last month?
e. Given that a customer uses paperless billing, what is the probability that the customer churned in the last month?
f. Given that a customer did not use paperless billing, what is the probability the customer churned in the last month?
g. What do the results of parts (e) and (f) tell you about the difference between those who use and do not use paperless billing concerning whether they churned in the last month?

Answers to Problems

1. a. 802/7,198 = 0.1114.
    b. 451/7,198 = 0.0627.
    c. 3,556/7,198 + 802/7,198 − 451/7,198 = 3,907/7,198 = 0.5428.
    d. (351/3,642)(350/3,641) = 0.0093.
2. a. 145/220 = 0.6591.
    b. 93/220 = 0.4227.
    c. 74/220 = 0.3364.
    d. 145/220 + 93/220 − 74/220 = 164/220 = 0.7455.
    e. 74/93 = 0.7957.
    f. 71/127 = 0.5591.
    g. Females are more likely to choose the local pizza.
3. a. 1,756/5,517 = 0.3183.
    b. 3,725/5,517 = 0.6752.
    c. 1,358/5,517 = 0.2461.
    d. 3,725/5,517 + 1,756/5,517 − 1,358/5,517 = 4,123/5,517 = 0.7473.
    e. 1,358/1,725 = 0.7872.
    f. 398/1,792 = 0.2210.
    g. Those who use paperless billing are much more likely to churn.

References

1. Berenson, M. L., D. M. Levine, K. A. Szabat, and D. Stephan. *Basic Business Statistics: Concepts and Applications*, 15th edition. Hoboken, NJ: Pearson Education, 2023.

2. Levine, D. M., D. Stephan, and K. A. Szabat. *Statistics for Managers Using Microsoft Excel*, 9th edition. Hoboken, NJ: Pearson Education, 2021.

CHAPTER 5

Probability Distributions

In Chapter 4, you learned the rules used to calculate the chance that a particular event will occur. In many situations, you can use a probability distribution to estimate the probability that particular events will occur.

5.1 Probability Distributions for Discrete Variables

A probability distribution for a variable summarizes or models the probabilities associated with the events for that variable. The form the distribution takes depends on whether the variable is discrete or continuous.

This section reviews probability distributions for discrete variables and statistics related to these distributions.

Discrete Probability Distribution

CONCEPT   A list of all possible distinct (elementary) events for a variable and their probabilities of occurrence.

EXAMPLE   See WORKED-OUT PROBLEM 1.

INTERPRETATION   In a probability distribution, the sum of the probabilities of all the events always equals 1. This is a way of saying that the (elementary) events listed are always collectively exhaustive—that is, that one of them must occur. Although you can use a table of outcomes to develop a probability distribution (see WORKED-OUT PROBLEM 2), you can also calculate probabilities for certain types of variables by using a formula that mathematically models the distribution.

WORKED-OUT PROBLEM 1   You want to determine the probability of getting 0, 1, 2, or 3 heads when you toss a fair coin (one with an equal probability of a head or a tail) three times in a row. Because getting 0, 1, 2, or 3 heads represent all possible distinct outcomes, you form a table of all possible outcomes (eight) of tossing a fair coin three times as follows.

Outcome	First Toss	Second Toss	Third Toss
1	Head	Head	Head
2	Head	Head	Tail
3	Head	Tail	Head
4	Head	Tail	Tail
5	Tail	Head	Head
6	Tail	Head	Tail
7	Tail	Tail	Head
8	Tail	Tail	Tail

From this table of all eight possible outcomes, you can form the summary table shown in Table 5.1.

From this probability distribution, you can determine that the probability of tossing three heads in a row is 0.125 and that the sum of the probabilities is 1.0, as it should be for a distribution of a discrete variable.

TABLE 5.1
Probability
Distribution
for Tossing
a Fair Coin
Three Times

Number of Heads	Number of Outcomes with That Number of Heads	Probability
0	1	1/8 = 0.125
1	3	3/8 = 0.375
2	3	3/8 = 0.375
3	1	1/8 = 0.125

Another way to compute these probabilities is to extend Rule 7 on page 80, the multiplication rule, to three events (or tosses). To get the probability of three heads, which is equal to 1/8 or 0.125 using Rule 7, you have:

$$P(H_1 \text{ and } H_2 \text{ and } H_3) = P(H_1) \times P(H_2) \times P(H_3)$$

Because the probability of heads in each toss is 0.5:

$$P(H_1 \text{ and } H_2 \text{ and } H_3) = (0.5)(0.5)(0.5)$$
$$P(H_1 \text{ and } H_2 \text{ and } H_3) = 0.125$$

The Expected Value of a Variable

CONCEPT   The sum of the products formed by multiplying each possible event in a discrete probability distribution by its corresponding probability.

INTERPRETATION   The expected value tells you the value of the variable that you could expect in the "long run"—that is, after many experimental trials. The expected value of a variable is also the mean (μ) of a variable.

WORKED-OUT PROBLEM 2   If there are three tosses of a coin (refer to Table 5.1), you can calculate the expected value of the number of heads as shown in Table 5.2:

Expected or Mean Value = Sum of [Each Value × The Probability of Each Value]

TABLE 5.2
Computing
the Expected
Value or
Mean of a
Probability
Distribution

Number of Heads	Probability	(Number of Heads) × (Probability)
0	0.125	(0) × (0.125) = 0
1	0.375	(1) × (0.375) = 0.375
2	0.375	(2) × (0.375) = 0.75
3	0.125	(3) × (0.125) = 0.125
		Expected or Mean Value = 1.50

Expected or Mean Value = (0)(0.125) + (1)(0.375) + (2)(0.375) + (3)(0.125)

= 0 + 0.375 + 0.750 + 0.375 = 1.50

Notice that in this example, the mean or expected value of the number of heads is 1.5, a value for the number of heads that is impossible. The mean of 1.5 heads tells you that, in the long run, if you toss three fair coins many times, the mean number of heads you can expect is 1.5.

Standard Deviation of a Variable (σ)

CONCEPT   The measure of variation around the expected value of a variable. You calculate this by first multiplying the squared difference between each value and the expected value by its corresponding probability. You then sum these products and then take the square root of that sum.

WORKED-OUT PROBLEM 3   If there are three tosses of a coin (refer to Table 5.1), you can calculate the variance and standard deviation of the number of heads, as shown in Table 5.3.

TABLE 5.3
Computing the Variance and Standard Deviation of a Probability Distribution

Number of Heads	Probability	(Number of Heads – Mean Number of Heads)^2 × (Probability)
0	0.125	$(0 – 1.5)^2$ × (0.125) = 2.25 × (0.125) = 0.28125
1	0.375	$(1 – 1.5)^2$ × (0.375) = 0.25 × (0.375) = 0.09375
2	0.375	$(2 – 1.5)^2$ × (0.375) = 0.25 × (0.375) = 0.09375
3	0.125	$(3 – 1.5)^2$ × (0.125) = 2.25 × (0.125) = 0.28125
		Total (Variance) = 0.75

σ = Square root of [Sum of (Squared differences between each value and the mean) × (Probability of the value)]

$$\sigma = \sqrt{(0-1.5)^2(0.125)+(1-1.5)^2(0.375)+(2-1.5)^2(0.375)+(3-1.5)^2(0.125)}$$
$$= \sqrt{2.25(0.125)+0.25(0.375)+0.25(0.375)+2.25(0.125)}$$
$$= \sqrt{0.75}$$

and

$$\sigma = \sqrt{0.75} = 0.866$$

INTERPRETATION   In financial analysis, you can use the standard deviation to assess the degree of risk of an investment, as WORKED-OUT PROBLEM 4 illustrates.

WORKED-OUT PROBLEM 4   Suppose that you are deciding between two alternative investments. Investment A is a mutual fund whose portfolio consists of a combination of stocks that make up the Dow Jones Industrial Average. Investment B consists of shares of a growth stock. You estimate the returns (per $1,000 investment) for each investment alternative under three economic condition events (recession, stable economy, and expanding economy) and also provide your subjective probability of the occurrence of each economic condition as follows.

Estimated Return for Two Investments Under Three Economic Conditions

		Investment	
Probability	Economic Event	Dow Jones Fund (A)	Growth Stock (B)
0.2	Recession	–$100	–$200
0.5	Stable economy	+100	+50
0.3	Expanding economy	+250	+350

Using the estimated returns, the following calculates the mean or expected return for the two investments:

$$\text{Mean} = \text{Sum of [Each value} \times \text{the probability of each value]}$$
$$\text{Dow Jones Fund mean} = (-100)(0.2) + (100)(0.5) + (250)(0.3) = \$105$$
$$\text{Growth stock mean} = (-200)(0.2) + (50)(0.5) + (350)(0.3) = \$90$$

You can calculate the standard deviation for the two investments, as shown in Tables 5.4 and 5.5.

TABLE 5.4
Computing the Variance and Standard Deviation for Dow Jones Fund (A)

Probability	Economic Event	Dow Jones Fund (A)	(Return – Mean Return)^2 × Probability
0.2	Recession	–$100	$(-100 - 105)^2 \times (0.2) = (42,025) \times (0.2) = 8,405$
0.5	Stable economy	+100	$(100 - 105)^2 \times (0.5) = (25) \times (0.5) = 12.5$
0.3	Expanding economy	+250	$(250 - 105)^2 \times (0.3) = (21,025) \times (0.3) = 6,307.5$
			Total (Variance) = 14,725

TABLE 5.5
Computing the Variance and Standard Deviation for Growth Stock (B)

Probability	Economic Event	Growth Stock (B)	(Return – Mean Return)^2 × Probability
0.2	Recession	–$200	$(-200 - 90)^2 \times (0.2) = (84,100) \times (0.2) = 16,820$
0.5	Stable economy	+50	$(50 - 90)^2 \times (0.5) = (1,600) \times (0.5) = 800$
0.3	Expanding economy	+350	$(350 - 90)^2 \times (0.3) = (67,600) \times (0.3) = 20,280$
			Total (Variance) = 37,900

$$\sigma = \text{Square root of [Sum of (Squared differences between a value and the mean)} \times \text{(Probability of the value)]}$$

$$\sigma_A = \sqrt{(-100-105)^2\,(0.2)+(100-105)^2\,(0.5)+(250-105)^2\,(0.3)}$$
$$= \sqrt{14{,}725}$$
$$= \$121.35$$

$$\sigma_B = \sqrt{(-200-90)^2\,(0.2)+(50-90)^2\,(0.5)+(350-90)^2\,(0.3)}$$
$$= \sqrt{37{,}900}$$
$$= \$194.68$$

The Dow Jones fund has a higher mean return than the growth fund and also has a lower standard deviation, indicating less variation in the return under the different economic conditions. Having a higher mean return with less variation makes the Dow Jones fund a more desirable investment than the growth fund.

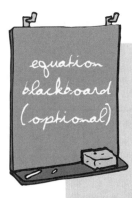

equation
blackboard
(optional)

interested
in
math?

To write the equations for the mean and standard deviation for a discrete probability distribution, you need the following symbols:

- An uppercase italic X, X, that represents a variable.

- An uppercase italic X with an italic lowercase i subscript, X_i, that represents the ith event associated with variable X.

- An uppercase italic N, N, that represents the number of elementary events for the variable X. (In Chapter 3, this symbol was used for the population size.)

- The symbol $P(X_i)$, which represents the probability of the event X_i.

- The symbol for the population mean, μ.

- The symbol for the population standard deviation, σ.

Using these symbols, you create these equations:

The mean of a probability distribution:

$$\mu = \sum_{i=1}^{N} X_i P(X_i)$$

The standard deviation of a probability distribution:

$$\sigma = \sqrt{\sum_{i=1}^{N}(X_i - \mu)^2\,P(X_i)}$$

5.2 The Binomial and Poisson Probability Distributions

As mentioned in the previous section, probability distributions for certain types of discrete variables can be modeled using a mathematical formula. This section looks at two important discrete distributions that are widely used to compute probabilities. The first probability distribution, the binomial, is used for variables that have only two mutually exclusive events. The second probability distribution, the Poisson, is used when you are counting the number of outcomes that occur in a unit.

The Binomial Distribution

important point ✏️

CONCEPT   The probability distribution for a discrete variable that meets these criteria:

- The variable is for a sample that consists of a fixed number of experimental trials (the sample size).
- The variable has only two mutually exclusive and collectively exhaustive events, typically labeled as success and failure.
- The probability of an event being classified as a success, p, and the probability of an event being classified as a failure, $1 - p$, are both constant in all experimental trials.
- The event (success or failure) of any single experimental trial is independent of (not influenced by) the event of any other trial.

EXAMPLE   The coin-tossing experiment described in WORKED-OUT PROBLEM 1.

INTERPRETATION   Using the binomial distribution prevents you from having to develop the probability distribution by using a table of outcomes and applying the multiplication rule—the method Section 4.3 uses. This distribution also does not require that the probability of success is 0.5, thereby enabling you to use it in more situations than the method that Section 5.1 discusses.

You can determine binomial probabilities by either using the formula that the next EQUATION BLACKBOARD presents, using a table of binomial probabilities, or using spreadsheet functions that create customized tables—the method WORKED-OUT PROBLEM 5, later in this chapter, uses.

Using spreadsheet functions or other software is the most typical way of calculating binomial probabilities, save for the simplest of problems in which the probability of success is 0.5. For those problems, you can use the table and multiplication rule method that Table 5.1 demonstrates. Using the results of that table, the probability of 0 heads is 0.125, the probability of 1 head is 0.375, the probability of 2 heads is 0.375, and the probability of 3 heads is 0.125 for the coin-tossing experiment.

Binomial distributions can be symmetrical or skewed. Whenever $p = 0.5$, the binomial distribution will be symmetrical, regardless of how large or small the value of the sample size, n. However, when $p \neq 0.5$, the distribution will be skewed. If $p < 0.5$, the distribution will be positive, or right-skewed; if $p > 0.5$, the distribution will be negative, or left-skewed. The distribution will become more symmetrical as p gets close to 0.5 and as the sample size, n, gets large.

The characteristics of the binomial distribution are:

Mean	The sample size (n) times the probability of success (p), $n \times p$, remembering that the sample size is the number of experimental trials.
Variance	The product of the sample size, probability of success, and probability of failure (1 – probability of success), $n \times p \times (1 - p)$.
Standard deviation	The square root of the variance, $\sqrt{np(1-p)}$.

WORKED-OUT PROBLEM 5   An online social networking website defines success if a web surfer stays and views its website for more than three minutes. Suppose that the probability that the surfer does stay for more than three minutes is 0.16. What is the probability that at least four (either four or five) of the next five surfers will stay for more than three minutes?

You need to sum the probabilities of four surfers staying and five surfers staying in order to determine the probabilities that at least four surfers stay.

These sums can be found in the $X = 4$ and $X = 5$ rows in the Binomial Probabilities Table of the **Chapter 5 Binomial** spreadsheet. From this table:

$$P(X = 4 \mid n = 5, p = 0.16) = 0.0028$$
$$P(X = 5 \mid n = 5, p = 0.16) = 0.0001$$

Therefore, the probability of four or more surfers staying and viewing the social networking website for more than three minutes is 0.0029 (which you compute by adding 0.0028 and 0.0001), or 0.29%.

spreadsheet solution

Binomial Probabilities

Chapter 5 Binomial contains a spreadsheet that calculates binomial probabilities for WORKED-OUT PROBLEM 5.

Best Practices

Use the **BINOM.DIST**(*number of successes, sample size, probability of success, cumulative*) function to calculate binomial probabilities.

How-Tos

For other problems, change the sample size in cell B4 and the probability of success in cell B5.

To have the Binomial Probabilities Table calculate the probability for exactly the *number of successes*, edit the BINOM.DIST functions in column B, changing *cumulative* from **False** to **True**.

	A	B	
1	Binomial Probabilities		
2			
3	Data		
4	Sample size	5	
5	Probability of success	0.16	
6			
7	Statistics		
8	Mean	0.8	=B4 * B5
9	Variance	0.672	=B8 * (1 - B5)
10	Standard deviation	0.8198	=SQRT(B9)
11			
12	Binomial Probabilities Table		
13		X	P(X)	
14		0	0.4182	=BINOM.DIST(A14, $B$4, $B$5, FALSE)
15		1	0.3983	=BINOM.DIST(A15, $B$4, $B$5, FALSE)
16		2	0.1517	=BINOM.DIST(A16, $B$4, $B$5, FALSE)
17		3	0.0289	=BINOM.DIST(A17, $B$4, $B$5, FALSE)
18		4	0.0028	=BINOM.DIST(A18, $B$4, $B$5, FALSE)
19		5	0.0001	=BINOM.DIST(A19, $B$4, $B$5, FALSE)

For the equation for the binomial distribution, use the symbols X (variable), n (sample size), and p (probability of success) previously introduced and add these symbols:

- A lowercase italic X, x, which represents the number of successes in the sample.

- The symbol $P(X = x \mid n, p)$, which represents the probability of the value x, given sample size n and probability of success p.

You use these symbols to form two separate expressions. One expression represents the number of ways you can get a certain number of successes in a certain number of trials:

$$\frac{n!}{x!(n-x)!}$$

(The symbol ! means factorial, where $n! = (n)(n-1)\ldots(1)$ so that 3! equals 6, $3 \times 2 \times 1$. 1! equals 1, and 0! is defined as being equal to 1.)

The second expression represents the probability of getting a certain number of successes in a certain number of trials *in a specific order*:

$$p^x \times (1-p)^{n-x}$$

Using these expressions, you form the following equation:

$$P(X = x \mid n, p) = \frac{n!}{x!(n-x)!} p^x (1-p)^{n-x}$$

As an example, the calculations for determining the binomial probability of 1 head in 3 tosses of a fair coin (that is, for a problem in which $n = 3$, $p = 0.5$, and $x = 1$) are as follows:

$$P(X = 1 \mid n = 3, p = 0.5) = \frac{3!}{1!(3-1)!}(0.5)^1 (1-0.5)^{3-1}$$

$$= \frac{3!}{1!(2)!}(0.5)^1 (1-0.5)^2$$

$$= 3(0.5)(0.25) = 0.375$$

Using symbols previously introduced, you can write the equation for the mean and standard deviation of the binomial distribution:

$$\mu = np$$

and:

$$\sigma = \sqrt{np(1-p)}$$

The Poisson Distribution

important point

CONCEPT The probability distribution for a discrete variable that meets these criteria:

- You are counting the number of times a particular event occurs in a unit.
- The probability that an event occurs in a particular unit is the same for all other units.
- The number of events that occur in a unit is independent of the number of events that occur in other units.
- As the unit gets smaller, the probability that two or more events will occur in that unit approaches zero.

EXAMPLES The number of computer network failures per day, the number of surface defects per square yard of floor covering, the number of customers arriving at a bank during the 12 noon to 1 p.m. hour, the number of fleas on the body of a dog.

INTERPRETATION To use the Poisson distribution, you define an area of opportunity, which is a continuous unit of area, time, or volume in which more than one event can occur. The Poisson distribution can model many variables that count the number of defects per area of opportunity or count the number of times items are processed from a waiting line.

You determine Poisson probabilities by applying the formula in the next EQUATION BLACKBOARD, by using a table of Poisson values, or by using functions that create customized tables—the method WORKED_OUT PROBLEM 6 uses. The characteristics of the Poisson distribution are:

Mean	The population mean, λ.
Variance	In the Poisson distribution, the variance is equal to the population mean, λ.
Standard deviation	The square root of the variance, or $\sqrt{\lambda}$.

spreadsheet solution

Poisson Probabilities

Chapter 5 Poisson contains a spreadsheet that calculates Poisson probabilities for WORKED-OUT PROBLEM 6.

Best Practices

Use the **POISSON.DIST**(*number of successes, mean number of successes, cumulative*) function to calculate Poisson probabilities.

How-Tos

For other problems, change the mean (expected) number of successes in cell B4.

To have the Poisson Probabilities Table calculate the probability for exactly the *number of successes*, edit the POISSON.DIST functions in column B, changing *cumulative* from **False** to **True**.

	A	B
1	**Poisson Probabilities**	
2		
3	**Data**	
4	**Mean (expected) number of successes**	**3**
5		
6	**Poisson Probabilities Table**	
7	**X**	**P(X)**
8	0	0.0498
9	1	0.1494
10	2	0.2240
11	3	0.2240
12	4	0.1680
13	5	0.1008
14	6	0.0504
15	7	0.0216
16	8	0.0081
17	9	0.0027
18	10	0.0008
19	11	0.0002
20	12	0.0001
21	13	0.0000
22	14	0.0000
23	15	0.0000

equation blackboard (optional)

interested in math?

For the equation for the Poisson distribution, use the symbols X (variable), n (sample size), and p (probability of success) previously introduced and add these symbols:

- A lowercase italic E, e, which represents the mathematical constant approximated by the value 2.71828.

- A lowercase Greek symbol lambda, λ, which represents the mean number of times that the event occurs per area of opportunity.

- A lowercase italic X, x, which represents the number of times the event occurs per area of opportunity.

- The symbol $P(X = x \mid \lambda)$, which represents the probability of x, given λ.

Using these symbols, you form the following equation:

$$P(X = x \mid \lambda) = \frac{e^{-\lambda}\lambda^{x}}{x!}$$

As an example, the calculations for determining the Poisson probability of exactly two arrivals in the next minute, given a mean of three arrivals per minute, is as follows:

$$P(X = 2 \mid \lambda = 3) = \frac{e^{-3}(3)^{2}}{2!}$$

$$= \frac{(2.71828)^{-3}(3)^{2}}{2!}$$

$$= \frac{(0.049787)(9)}{(2)}$$

$$= 0.2240$$

WORKED-OUT PROBLEM 6   You want to determine the probabilities that a specific number of customers will arrive at a bank branch in a one-minute interval during the lunch hour: Will zero customers arrive, one customer, two customers, and so on? You determine that you can use the Poisson distribution for the following reasons:

- The variable is a count per unit—that is, customers per minute.
- You assume that the probability that a customer arrives during a specific one-minute interval is the same as the probability for all the other one-minute intervals.
- Each customer's arrival has no effect on (is independent of) all other arrivals.
- The probability that two or more customers will arrive in a given time period approaches zero as the time interval decreases from one minute.

Using historical data, you determine that the mean number of arrivals of customers is three per minute during the lunch hour. Using the **Chapter 5 Poisson** spreadsheet, in which 3 has already been entered as the value for mean (expected) number of successes, you determine that for this problem:

- The probability of zero arrivals is 0.0498.
- The probability of one arrival is 0.1494.
- The probability of two arrivals is 0.2240.

Therefore, the probability of two or fewer customer arrivals per minute at the bank during the lunch hour is 0.4232, the sum of the probabilities for zero, one, and two arrivals (0.0498 + 0.1494 + 0.2240 = 0.4232).

5.3 Continuous Probability Distributions and the Normal Distribution

Probability distributions can also be developed to model continuous variables. The exact mathematical expression for the probability distribution for a continuous variable involves integral calculus and is not shown in this book.

Probability Distribution for a Continuous Variable

CONCEPT   The area under a curve that represents the probabilities for a continuous variable.

EXAMPLE   See the example for the normal distribution.

important point

INTERPRETATION   Probability distributions for a continuous variable differ from discrete distributions in several important ways:

- An event can take on any value within the range of the variable and not just an integer value.
- The probability of any specific value is zero.
- Probabilities are expressed in terms of an area under a curve that represents the continuous distribution.

One continuous distribution, the **normal distribution**, is especially important in statistics because it can model many different continuous variables.

Normal Distribution

important point

CONCEPT   The probability distribution for a continuous variable that meets these criteria:

- The graphed curve of the distribution is bell-shaped and symmetrical.
- The mean, median, and mode are all the same value.
- The population mean, μ, and the population standard deviation, σ, determine probabilities.
- The distribution extends from negative to positive infinity. (The distribution has an infinite range.)
- Probabilities are always cumulative and expressed as inequalities, such as $P < X$ or $P \geq X$, where X is a value for the variable.

EXAMPLE   The normal distribution appears as a bell-shaped curve, as the following figure illustrates.

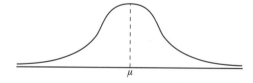

μ

INTERPRETATION   The importance of the normal distribution to statistics cannot be overstated. Probabilities associated with variables as diverse as physical characteristics such as height and weight, scores on standardized exams, and the dimension of industrial parts, tend to follow a normal distribution. Under certain circumstances, the normal distribution also approximates various discrete probability distributions, such as the binomial and Poisson distributions. In addition, the normal distribution provides the basis for classical statistical inference that Chapters 6 through 9 discuss.

You determine normal probabilities by using a table of normal probabilities (such as Table C.1 in Appendix C) or by using software functions. Normal probability tables such as Table C.1 and some software functions use a standardized normal distribution that requires you to convert an X value of a variable to its corresponding Z score (Section 3.3). You perform this conversion by subtracting the population mean μ from the X value and dividing the resulting difference by the population standard deviation σ, expressed algebraically as follows:

important point

$$Z = \frac{X - \mu}{\sigma}$$

When the mean is 0 and the standard deviation is 1, the X value and Z score will be the same, and no conversion is necessary.

WORKED-OUT PROBLEM 7   Packages of chocolate candies have a labeled weight of 6 ounces. In order to ensure that very few packages have a weight below 6 ounces, the filling process provides a mean weight above 6 ounces. In the past, the mean weight has been 6.15 ounces with a standard deviation of 0.05 ounce. Suppose you want to determine the probability that a single package of chocolate candies will weigh between 6.15 and 6.20 ounces. To determine this probability, you use Table C.1, the table of the probabilities of the cumulative standardized normal distribution.

To use Table C.1, you must first convert the weights to Z scores by subtracting the mean and then dividing by the standard deviation, as shown here:

$$Z(\text{lower}) = \frac{6.15 - 6.15}{0.05} = 0 \quad Z(\text{upper}) = \frac{6.20 - 6.15}{0.05} = 1.0$$

As the next step, you need to determine the probability that corresponds to the area between 0 and +1 Z units (standard deviations). To do this, you take the cumulative probability associated with 0 Z units and subtract it from the cumulative probability associated with +1 Z units. Using Table C.1, you determine that these probabilities are 0.8413 and 0.5000, as the table on the next page illustrates.

Finding a Cumulative Area Under the Normal Curve

					Cumulative Probabilities				
Z	**.00**	**.01**	**.02**	**.03**	**.04**	**.05**	**.06**	**.07**	**.08**	**.09**
0.0	.5000	.5040	.5080	.5120	.5160	.5199	.5239	.5279	.5319	.5359
0.1	.5398	.5438	.5478	.5517	.5557	.5596	.5636	.5675	.5714	.5753
0.2	.5793	.5832	.5871	.5910	.5948	.5987	.6026	.6064	.6103	.6141
0.3	.6179	.6217	.6255	.6293	.6331	.6368	.6406	.6443	.6480	.6517
0.4	.6554	.6591	.6628	.6664	.6700	.6736	.6772	.6808	.6844	.6879
0.5	.6915	.6950	.6985	.7019	.7054	.7088	.7123	.7157	.7190	.7224
0.6	.7257	.7291	.7324	.7357	.7389	.7422	.7454	.7486	.7518	.7549
0.7	.7580	.7612	.7642	.7673	.7704	.7734	.7764	.7794	.7823	.7852
0.8	.7881	.7910	.7939	.7967	.7995	.8023	.8051	.8078	.8106	.8133
0.9	.8159	.8186	.8212	.8238	.8264	.8289	.8315	.8340	.8365	.8389
1.0	.8413	.8438	.8461	.8485	.8508	.8531	.8554	.8577	.8599	.8621

Therefore, the probability that a single package of chocolate candies will weigh between 6.15 and 6.20 ounces is 0.3413 (0.8413 − 0.5000 = 0.3413). Using the cumulative probability associated with +1 Z units, you can also calculate that the probability that a package weighs more than 6.20 ounces is 0.1587 (1 − 0.8413). (The **Chapter 5 Normal** worksheet that the next Spreadsheet Solution discusses can also be used to produce these results.) The following chart visualizes the calculations of WORKED-OUT PROBLEM 7.

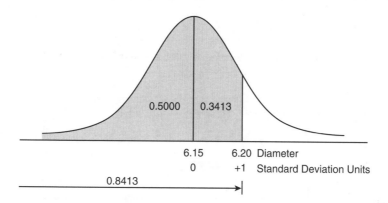

Using Standard Deviation Units

Because of the equivalence between Z scores and standard deviation units, probabilities of the normal distribution are often expressed as ranges of plus-or-minus standard deviation units. Such probabilities can be determined directly from Table C.1, the table of the probabilities of the cumulative standardized normal distribution.

For example, to determine the normal probability associated with the range ±3 standard deviations, you would use Table C.1 to look up the probabilities associated with $Z = -3.00$ and $Z = +3.00$:

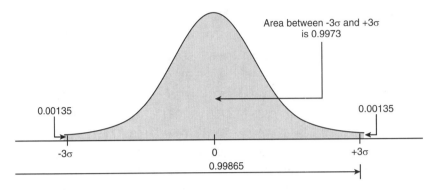

Table 5.6 represents the appropriate portion of Table C.1 for $Z = -3.00$. From this table excerpt, you can determine that the probability of a value less than $Z = -3$ units is 0.00135.

TABLE 5.6

Partial Table C.1 for Obtaining a Cumulative Area Below −3 Z Units

Z	.00	.01	.02	.03	.04	.05	.06	.07	.08	.09
.		.	.	.	.	.	.	.	.	.
.		.	.	.	.	.	.	.	.	.
.		.	.	.	.	.	.	.	.	.
−3.0	0.00135	0.00131	0.00126	0.00122	0.00118	0.00114	0.00111	0.00107	0.00103	0.00100

Table 5.7 represents the portion of Table C.1 for $Z = +3.00$. From this table excerpt, you can determine that the probability of a value less than $Z = +3$ units is 0.99865.

TABLE 5.7

Partial Table C.1 for Obtaining a Cumulative Area Below +3 Z Units

Z	.00	.01	.02	.03	.04	.05	.06	.07	.08	.09
.		.	.	.	.	.	.	.	.	.
.		.	.	.	.	.	.	.	.	.
.		.	.	.	.	.	.	.	.	.
+3.0	0.99865	0.99869	0.99874	0.99878	0.99882	0.99886	0.99889	0.99893	0.99897	0.99900

Therefore, the probability associated with the range plus-or-minus 3 standard deviations in a normal distribution is 0.9973 (0.99865 − 0.00135). Stated another way, the probability is 0.0027 (2.7 out of a thousand chance) that a value will not be within the range of plus-or-minus 3 standard deviations. Table 5.8 summarizes probabilities for several different ranges of standard deviation units.

TABLE 5.8
Probabilities for Different Standard Deviation Ranges

Standard Deviation Unit Ranges	Probability or Area Outside These Units	Probability or Area Within These Units
−1σ to +1σ	0.3174	0.6826
−2σ to +2σ	0.0455	0.9545
−3σ to +3σ	0.0027	0.9973
−6σ to +6σ	0.000000002	0.999999998

Finding the Z Value from the Area Under the Normal Curve

Each of the previous examples involved using the normal distribution table to find an area under the normal curve that corresponded to a specific Z value. Sometimes you need to do the opposite of this and find the Z value that corresponds to a specific area. For example, you might want to find the Z value that corresponds to a cumulative area of 1%, 5%, 95%, or 99%. You might also want to find the lower and upper Z values between which 95% of the area under the curve is contained.

To find the Z value that corresponds to a cumulative area, you locate the cumulative area in the body of the normal table, or the closest value to the cumulative area you want to find, and then determine the Z value that corresponds to this cumulative area.

WORKED-OUT PROBLEM 8   You want to find the Z values such that 95% of the normal curve is contained between a lower Z value and an upper Z value, with 2.5% below the lower Z value and 2.5% above the upper Z value.

Using the figure on page 106, you determine that you need to find the Z value that corresponds to a cumulative area of 0.025 and the Z value that corresponds to a cumulative area of 0.975.

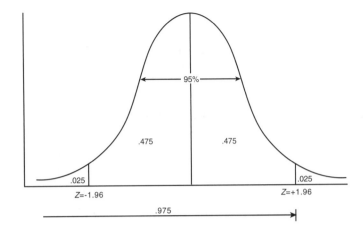

Table 5.9 contains a portion of Table C.1 that is needed to find the Z value that corresponds to a cumulative area of 0.025. Table 5.10 contains a portion of Table C.1 that is needed to find the Z value that corresponds to a cumulative area of 0.975.

TABLE 5.9

Partial Table C.1 for Finding Z Value That Corresponds to a Cumulative Area of 0.025

Z	.00	.01	.02	.03	.04	.05	.06	.07	.08	.09
.	.	.	.	.	.	.		.	.	.
.	.	.	.	.	.	.		.	.	.
.	.	.	.	.	.	.	.	.	.	.
−2.0	0.0228	0.0222	0.0217	0.0212	0.0207	0.0202	0.0197	0.0192	0.0188	0.0183
−1.9	0.0287	0.0281	0.0274	0.0268	0.0262	0.0256	0.0250	0.0244	0.0239	0.0233

TABLE 5.10

Partial Table C.1 for Finding Z Value That Corresponds to a Cumulative Area of 0.975

Z	.00	.01	.02	.03	.04	.05	.06	.07	.08	.09
.	.	.	.	.	.	.		.	.	.
.	.	.	.	.	.	.		.	.	.
.	.	.	.	.	.	.	.	.	.	.
1.9	0.9713	0.9719	0.9726	0.9732	0.9738	0.9744	0.9750	0.9756	0.9761	0.9767
2.0	0.9772	0.9778	0.9783	0.9788	0.9793	0.9798	0.9803	0.9808	0.9812	0.9817

To find the Z value that corresponds to a cumulative area of 0.025, you look in the body of Table 5.9 until you see the value 0.025. Then you determine the row and column that this value corresponds to. Locating the value of 0.025, you

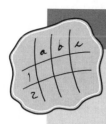

spreadsheet solution

Normal Probabilities

Chapter 5 Normal contains a spreadsheet that calculates various normal probabilities.

Best Practices

Use the **STANDARDIZE, NORM.DIST, NORM.S.INV**, and **NORM.INV** functions to calculate values associated with normal probabilities.

How-Tos

For other problems, first change the Common Data, the mean in cell B4 and the standard deviation in cell B5, and then supply other values as necessary.

Function Tip FT2 in Appendix D explains how to enter the functions used in **Chapter 5 Normal**.

Advanced Technique ADV3 in Appendix E explains how the spreadsheet uses the ampersand operator (&) to produce some of the text labels in columns A and D.

▲	A	B	C	D	E
1	Normal Probabilities				
2					
3	Common Data			Probability for a Range	
4	Mean	6.15		From X Value	6.15
5	Standard Deviation	0.05		To X Value	6.2
6				Z Value for 6.15	0
7	Probability for X <=			Z Value for 6.2	1
8	X Value	6.2		P(X<=6.15)	0.5000
9	Z Value	1		P(X<=6.2)	0.8413
10	P(X<=6.2)	0.8413		P(6.15<=X<=6.2)	0.3413
11					
12	Probability for X >			Find Z and X Given a Cumulative Pctage.	
13	X Value	6.2		Cumulative Percentage	10.00%
14	Z Value	1		Z Value	-1.28
15	P(X>6.2)	0.1587		X Value	6.09
16					
17				Find Z and X Values Given a Percentage	
18				Percentage	95.00%
19				Z Value	-1.96
20				Lower X Value	6.05
21				Upper X Value	6.25

see that it is located in the –1.9 row and the .06 column. Thus, the Z value that corresponds to a cumulative area of 0.025 is –1.96.

To find the Z value that corresponds to a cumulative area of 0.975, you look in the body of Table 5.10 until you see the value 0.975. Then you determine the corresponding row and column that this value belongs to. Locating the value of 0.975, you see that it is in thé 1.9 row and the .06 column. Thus, the Z value that corresponds to a cumulative area of 0.975 is 1.96. Taking this result along with the Z value of –1.96 for a cumulative area of 0.025 means that 95% of all the values will be between $Z = -1.96$ and $Z = 1.96$.

WORKED-OUT PROBLEM 9   You want to find the weights that will contain 95% of the packages of chocolate candy that WORKED-OUT PROBLEM 7 discusses. To do this, you need to determine X in the formula:

$$Z = \frac{X - \mu}{\sigma}$$

Solving this formula for X, you have:

$$X = \mu + Z\sigma$$

Because the mean weight is 6.15 ounces and the standard deviation is 0.05 ounce, and 95% of the packages will be contained between –1.96 and +1.96 standard deviation (Z) units, the interval that will contain 95% of the packages will be between:

$$6.15 + (-1.96)(0.05) = 6.15 - 0.098 = 6.052 \text{ ounces}$$

and:

$$6.15 + (+1.96)(0.05) = 6.15 + 0.098 = 6.248 \text{ ounces}$$

5.4 The Normal Probability Plot

In order to use many inferential statistical methods, you must determine if a set of data approximately follows a normal distribution. One technique for making this determination is to use the **normal probability plot**.

CONCEPT   A graph that plots the relationship between ranked data values and the Z scores that these values would correspond to if the set of data values follows a normal distribution. If the data values follow a normal distribution, the graph will be linear (a straight line), as shown in the following example.

EXAMPLES

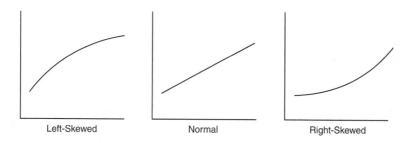

Left-Skewed   Normal   Right-Skewed

INTERPRETATION   Normal probability plots are based on the idea that the Z scores for the ranked values increase at a predictable rate for data that follow a normal distribution. The exact details to produce a normal probability plot can vary, but one common approach is to use a **quantile–quantile plot**. In this method, each value (ranked from lowest to highest) is plotted on the vertical, Y, axis, and its transformed Z score is plotted on the horizontal, X, axis. If the data are normally distributed, a plot of the data ranked from lowest to highest will follow a straight line. As shown in the preceding figure, if the data are left-skewed, the curve will rise more rapidly at first and then level off. If the data are right-skewed, the data will rise more slowly at first and then rise at a faster rate for higher values of the variable being plotted.

NBA Ticket Cost

WORKED-OUT PROBLEM 10   You seek to determine whether the data for NBA average ticket cost data, first used in the Chapter 3 WORKED-OUT PROBLEM 12, follows a normal distribution. Using Microsoft Excel, you create this normal probability plot:

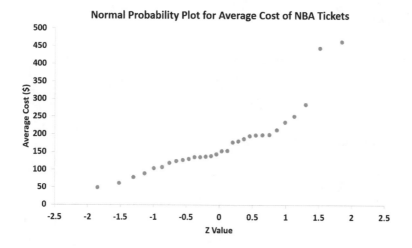

The data points in this normal probability plot form a curve, suggesting that the NBA average ticket cost data is right-skewed (and does not follow a normal distribution).

Important Equations

Mean of a Probability Distribution:

$$\mu = \sum_{i=1}^{N} X_i P(X_i)$$

Standard Deviation of a Discrete Probability Distribution:

$$\sigma = \sqrt{\sum_{i=1}^{N} (X_i - \mu)^2 P(X_i)}$$

Binomial Distribution:

$$P(X = x \mid n, p) = \frac{n!}{x!(n-x)!} p^x (1-p)^{n-x}$$

Mean of the Binomial Distribution:

$$\mu = np$$

Standard Deviation of the Binomial Distribution:

$$\sigma_x = \sqrt{np(1-p)}$$

Poisson Distribution:

$$P(X = x \mid \lambda) = \frac{e^{-\lambda} \lambda^x}{x!}$$

Normal Distribution—Finding a Z Value:

$$Z = \frac{X - \mu}{\sigma}$$

Normal Distribution—Finding an X Value:

$$X = \mu + Z\sigma$$

One-Minute Summary
Probability Distributions

Discrete probability distribution concepts:

- Expected value
- Variance σ^2 and standard deviation σ
- Is there a fixed sample size n, and is each observation classified into one of two categories?
  - If yes, use the binomial distribution, subject to other conditions.
  - If no, use the Poisson distribution, subject to other conditions.

Continuous probability distribution concepts:

- Normal distribution
- Normal probability plot

Test Yourself
Short Answers

1. The sum of the probabilities of all the events in a probability distribution is equal to:

   a.   0
   b.   the mean
   c.   the standard deviation
   d.   1

2. The largest number of possible successes in a binomial distribution is:

   a.   0
   b.   1
   c. n
   d.   infinite

3. The smallest number of possible successes in a binomial distribution is:

   a.   0
   b.   1
   c. n
   d.   infinite

4. Which of the following about the binomial distribution is not a true statement?

   a.   The probability of success must be constant from trial to trial.
   b.   Each outcome is independent of the other.

    c.   Each outcome may be classified as either "success" or "failure."

    d.   The variable of interest is continuous.

5. Whenever $p = 0.5$, the binomial distribution will:

    a.   always be symmetric

    b.   be symmetric only if n is large

    c.   be right-skewed

    d.   be left-skewed

6. What type of probability distribution will the consulting firm most likely employ to analyze the insurance claims in the following problem?

An insurance company has called a consulting firm to determine whether the company has an unusually high number of false insurance claims. It is known that the industry proportion for false claims is 6%. The consulting firm has decided to randomly and independently sample 50 of the company's insurance claims. They believe that the number of false claims from the sample will yield the information the company desires.

    a.   binomial distribution

    b.   Poisson distribution

    c.   normal distribution

    d.   none of the above

7. The service manager for a new automobile dealership reviewed dealership records of the past 20 sales of new cars to determine the number of warranty repairs he will be called on to perform in the next 30 days.

Corporate reports indicate that the probability any one of their new cars needs a warranty repair in the first 30 days is 0.035. The manager assumes that calls for warranty repair are independent of one another and is interested in predicting the number of warranty repairs he will be called on to perform in the next 30 days for this batch of 20 new cars sold.

What type of probability distribution will most likely be used to analyze warranty repair needs on new cars in the following problem?

    a.   binomial distribution

    b.   Poisson distribution

    c.   normal distribution

    d.   none of the above

8. The quality control manager of Marilyn's Cookies is inspecting a batch of chocolate chip cookies. When the production process is in control, the mean number of chocolate chip parts per cookie is 6.0. The manager is interested in analyzing the probability that any particular cookie being inspected has fewer than 10.0 chip parts. What probability distribution should be used?

    a.   binomial distribution

    b.   Poisson distribution

   c.   normal distribution

   d.   none of the above

9. The smallest number of possible successes in a Poisson distribution is:

   a.   0

   b.   1

   c. n

   d.   infinite

10. Based on past experience, the time you spend on emails per day has a mean of 30 minutes and a standard deviation of 10 minutes. To compute the probability of spending at least 12 minutes on emails, you use which probability distribution?

   a.   binomial distribution

   b.   Poisson distribution

   c.   normal distribution

   d.   none of the above

11. A computer lab at a university has ten personal computers. Based on past experience, the probability that any one of them will require repair on a given day is 0.05. To find the probability that exactly two of the computers will require repair on a given day, you use which probability distribution?

   a.   binomial distribution

   b.   Poisson distribution

   c.   normal distribution

   d.   none of the above

12. The mean number of customers who arrive per minute at any one of the checkout counters of a grocery store is 1.8. What probability distribution can be used to find out the probability that there will be no customers arriving at a checkout counter in the next minute?

   a.   binomial distribution

   b.   Poisson distribution

   c.   normal distribution

   d.   none of the above

13. A multiple-choice test has 25 questions. There are four choices for each question. A student who has not studied for the test decides to answer all questions by randomly choosing one of the four choices for each question.

What probability distribution can be used to compute his chance of correctly answering at least 15 questions?

   a.   binomial distribution

   b.   Poisson distribution

   c.   normal distribution

   d.   none of the above

14. Which of the following are true statements about the normal distribution?
    a.   Theoretically, the mean, median, and mode are the same.
    b.   About 99.7% of the values fall within 3 standard deviations of the mean.
    c.   It is defined by two characteristics, μ and σ.
    d.   All of the above are true.

15. Which of the following is not true of the normal distribution?
    a.   Theoretically, the mean, median, and mode are the same.
    b.   About two-thirds of the observations fall within 1 standard deviation of the mean.
    c.   It is a discrete probability distribution.
    d.   Its parameters are the mean, μ, and standard deviation, σ.

16. The probability that Z is less than −1.0 is _____ the probability that Z is greater than +1.0.
    a.   less than
    b.   the same as
    c.   greater than

17. The normal distribution is _____ in shape:
    a.   right-skewed
    b.   left-skewed
    c.   symmetric

18. If a particular set of data is approximately normally distributed, you would find that approximately:
    a.   2 of every 3 observations would fall between 1 standard deviation around the mean
    b.   4 of every 5 observations would fall between 1.28 standard deviations around the mean
    c.   19 of every 20 observations would fall between 2 standard deviations around the mean
    d.   all of the above

19. Given that X is a normally distributed variable with a mean of 50 and a standard deviation of 2, the probability that X is between 47 and 54 is

    _____.

Answer True or False:

20. Theoretically, the mean, median, and the mode are all equal for a normal distribution.

21. Another name for the mean of a probability distribution is the expected value.

22. The diameters of 100 randomly selected bolts follow a binomial distribution.

23. If the data values are normally distributed, the normal probability plot will follow a straight line.

Answers to Test Yourself Short Answers

1. d	13. a
2. c	14. d
3. a	15. c
4. d	16. b
5. a	17. c
6. a	18. d
7. a	19. 0.9104
8. b	20. True
9. a	21. True
10. c	22. False
11. a	23. True
12. b	

Problems

1. Given the following probability distributions:

Distribution A		Distribution B	
X	P(X)	X	P(X)
0	0.20	0	0.10
1	0.20	1	0.20
2	0.20	2	0.40
3	0.20	3	0.20
4	0.20	4	0.10

    a.   Compute the expected value of each distribution.
    b.   Compute the standard deviation of each distribution.
    c.   Compare the results of distributions A and B.

2. In the carnival game Under-or-Over-Seven, a pair of fair dice is rolled once, and the resulting sum determines whether the player wins or loses his or her bet. For example, the player can bet $1 that the sum will be under 7— that is, 2, 3, 4, 5, or 6. For this bet, the player wins $1 if the result is under 7 and loses $1 if the outcome equals or is greater than 7.

Similarly, the player can bet $1 that the sum will be over 7—that is, 8, 9, 10, 11, or 12. Here, the player wins $1 if the result is over 7 but loses $1 if the result is 7 or under. A third method of play is to bet $1 on the outcome 7. For this bet, the player wins $4 if the result of the roll is 7 and loses $1 otherwise.

a. Construct the probability distribution representing the different outcomes that are possible for a $1 bet on being under 7.

b. Construct the probability distribution representing the different outcomes that are possible for a $1 bet on being over 7.

c. Construct the probability distribution representing the different outcomes that are possible for a $1 bet on 7.

d. Show that the expected long-run profit (or loss) to the player is the same no matter which method of play is used.

3. The number of arrivals per minute at a bank located in the central business district of a large city was recorded over a period of 200 minutes with the following results:

Arrivals	Frequency
0	14
1	31
2	47
3	41
4	29
5	21
6	10
7	5
8	2

a. Compute the expected number of arrivals per day.

b. Compute the standard deviation.

4. Suppose that a judge's decisions are upheld by an appeals court 90% of the time. In her next ten decisions, what is the probability that

a. eight of her decisions are upheld by an appeals court?

b. all ten of her decisions are upheld by an appeals court?

c. eight or more of her decisions are upheld by an appeals court?

5. A venture capitalist firm that specializes in funding risky high-technology startup companies has determined that only one in ten of its companies is a "success" that makes a substantive profit within six years. Given this

historical record, what is the probability that in the next three startups it finances, the firm will have:

a. exactly one success?
b. exactly two successes?
c. fewer than two successes?
d. at least two successes?

6. Accuracy in taking orders at a drive-through window is important for fast-food chains. Periodically, *QSR Magazine* publishes the results of a survey that measures accuracy, defined as the percentage of orders that are filled correctly. In a recent month, the percentage of orders filled correctly at Burger King was approximately 90.9%. (Data extracted from "The Drive-Thru Performance Study: Order Accuracy," https://bit.ly/3ijbOfV.)

Suppose that you go to the drive-through window at Burger King and place an order. Two friends of yours independently place orders at the drive-through window at the same Burger King.

a. What is the probability that all three of the three orders will be filled correctly?
b. What is the probability that none of the three orders will be filled correctly?
c. What is the probability that at least two of the three orders will be filled correctly?
d. What are the mean and standard deviation of the binomial distribution for the number of orders filled correctly?

7. The number of power outages at a power plant has a Poisson distribution with a mean of four outages per year. What is the probability that in a year there will be

a. no power outages?
b. four power outages?
c. at least three power outages?

8. The quality control manager of Marilyn's Cookies is inspecting a batch of chocolate-chip cookies that has just been baked. When the production process is in control, the mean number of chip parts per cookie is 6.0. What is the probability that in any particular cookie being inspected there are

a. fewer than five chip parts?
b. exactly five chip parts?
c. five or more chip parts?
d. either four or five chip parts?

9. The U.S. Department of Transportation maintains statistics for mishandled bags per 1,000 airline passengers. In a recent month, domestic airlines had 5.82 mishandled bags per 1,000 passengers. What is the probability that in the next 1,000 passengers, domestic airlines will have

   a. no mishandled bags?
   b. at least one mishandled bag?
   c. at least two mishandled bags?

   In a recent month, European airlines had 7.21 mishandled bags per 1,000 passengers. What is the probability that in the next 1,000 passengers, European airlines will have

   d. no mishandled bags?
   e. at least one mishandled bag?
   f. at least two mishandled bags?

   In a recent month, Asian airlines had 1.69 mishandled bags per 1,000 passengers. What is the probability that in the next 1,000 passengers, Asian airlines will have

   g. no mishandled bags?
   h. at least one mishandled bag?
   i. at least two mishandled bags?
   j. Compare the results for domestic airlines to the results for European and Asian airlines.

10. Assume that X is a normally distributed variable with a mean of 50 and a standard deviation of 2.

    a. What is the probability that X is between 47 and 54?
    b. What is the probability that X is less than 55?
    c. There is a 90% chance that X will be less than what value?

11. A set of final examination grades in an introductory statistics course is normally distributed, with a mean of 73 and a standard deviation of 8.

    a. What is the probability of getting a grade below 91 on this exam?
    b. What is the probability that a student scored between 65 and 89?
    c. The probability is 5% that a student taking the test scores higher than what grade?
    d. If the professor grades on a curve (that is, gives A's to the top 10% of the class, regardless of the score), are you better off with a grade of 81 on this exam or a grade of 68 on a different exam, where the mean is 62 and the standard deviation is 3? Explain.

12. The owner of a fish market determined that the mean weight for salmon is 12.3 pounds, with a standard deviation of 2 pounds. Assume that the weights of salmon are normally distributed

    a.   What is the probability that a randomly selected salmon will weigh between 12 and 15 pounds?

    b.   What is the probability that a randomly selected salmon will weigh less than 10 pounds?

    c.   95% of the salmon will weigh between what two values?

13. The file **MPG** represents the miles per gallon for 2021 sedans. (Data extracted from "Cars," https://consumerreports.org/cars.) Does the miles per gallon for 2021 sedans follow a normal distribution?

MPG

14. The file **Domestic Beer** contains the percentage alcohol, number of calories per 12 ounces, and number of carbohydrates (in grams) per 12 ounces for 157 of the best-selling domestic beers in the United States. (Data extracted from "Find Out How Many Calories in Beer?" https://www.beer100.com/beer-calories.) Do you think any of these variables are normally distributed? Explain.

Domestic Beer

Answers to Problems

1. a. A: $\mu = 2$, B: $\mu = 2$.

  b. A: $\sigma = 1.414$, B: $\sigma = 1.095$.

  c. Distribution A is uniform and symmetric; Distribution B is symmetric and has a smaller standard deviation than distribution A.

2. a.

X	$P(X)$
$-1	21/36
$+1	15/36

  b.

X	$P(X)$
$-1	21/36
$+1	15/36

  c.

X	$P(X)$
$-1	30/36
$+1	6/36

  d. $-0.167 for each method of play.

3. a. 2.90   b. 1.772

4. a. 0.1937   b. 0.3487   c. 0.9298

5. a. 0.243   b. 0.027   c. 0.972   d. 0.028

6. a. 0.7511   b. 0.0008   c. 0.9767   d. 2.727 and 0.4982

7. a. 0.0183   b. 0.1954   c. 0.7619

8. a. 0.2851   b. 0.1606   c. 0.7149   d. 0.2945
9. a. 0.0030   b. 0.9970   c. 0.9797
    d. 0.0007   e. 0.9993   f. 0.9939
    g. 0.1845   h. 0.8155   i. 0.5036
    j. Mishandled bags are more likely in Europe and less likely in Asia than in the United States.
10. a. 0.9104   b. 0.9938   c. 52.5631
11. a. 0.9878   b. 0.8185   c. 86.16%
    d. Option 1: Because your score of 81% on this exam represents a Z score of 1.00, which is below the minimum Z score of 1.28, you will not earn an A grade on the exam under this grading option. Option 2: Because your score of 68% on this exam represents a Z score of 2.00, which is well above the minimum Z score of 1.28, you will earn an A grade on the exam under this grading option. You should prefer option 2.
12. a. 0.4711
    b. 0.1251
    c. 8.38 and 16.22
13. The average miles per gallon for 2021 sedans is approximately normally distributed because the normal probability plot is approximately a straight line.
14. The alcohol percentage, calories, and carbohydrates all appear to be right-skewed.

References

1. Berenson, M. L., D. M. Levine, K. A. Szabat, and D. Stephan. *Basic Business Statistics: Concepts and Applications*, 15th edition. Hoboken, NJ: Pearson Education, 2023.

2. Levine, D. M., D. Stephan, and K. A. Szabat. *Statistics for Managers Using Microsoft Excel*, 9th edition. Hoboken, NJ: Pearson Education, 2021.

3. Levine, D. M., P. P. Ramsey, and R. K. Smidt. *Applied Statistics for Engineers and Scientists Using Microsoft Excel and Minitab*. Upper Saddle River, NJ: Prentice Hall, 2001.

Sampling Distributions and Confidence Intervals

Inferential statistics relies on the principle that sample statistics can be used to reach conclusions about population parameters. The idea that a small part of a much larger thing, *a sample*, can be used to reach conclusions about that much larger thing, *a population*, seems counterintuitive to some who may be initially dismissive about inferential methods.

This chapter explains the validity of the principle on which inferential statistics relies. You will learn the concept of *statistical* confidence and the statistical methods that enable you to estimate a *confidence interval*.

6.1   Foundational Concepts

Learning several foundational concepts first helps you understand the principle that underlies inferential statistics. These concepts, when combined with the probability and probability distribution concepts of the previous two chapters, help explain the principle.

All Possible Samples of a Given Sample Size

CONCEPT   The set of samples that represents all possible combinations of values from a population for a specific sample size.

EXAMPLES   For a population size of 10, the set of all possible samples of $n = 2$ without replacement would be 45 samples.

INTERPRETATION   The size of the set of all possible samples of a given sample size can be calculated using combinational mathematics. The set can grow quite large as the population size or sample size increases. For example, if the population size was 100 instead of 10, the set of all possible samples of $n = 2$ without replacement would contain 4,950 different samples.

When examining all possible samples, you work with a sample statistic—such as the mean for numerical data and the proportion for categorical data—and not the samples themselves. This leads directly to the next concept.

Sampling Distribution

CONCEPT   The probability distribution of a sample statistic for all possible samples of a given size n for a given population.

EXAMPLES   The sampling distribution of the mean, the sampling distribution of the proportion.

INTERPRETATION   Sampling distributions express the variability in the sample statistic computed for each sample, just as a probability distribution expresses that variability for the values of a variable. For example, in the sampling distribution of the mean for a numerical variable, some of the sample means might be smaller than others, some might be larger, and some might be similar to each other. If you knew about this variability in the way that variability is "known" in the probability distributions that Chapter 5 discusses, you could use techniques related to those that Chapter 5 presents to express the likelihood that the sample statistic properly estimates a population parameter. Calculating the sample statistic for all possible samples would be one way but would be impractical for typical population sizes. What then could you do?

Central Limit Theorem

CONCEPT The sampling distribution of the mean can be approximated by the normal distribution when the sample size of all possible samples gets large enough.

INTERPRETATION The Central Limit Theorem (CLT) enables you to determine the shape of a sampling distribution without having to resort to the (impractical) method of calculating sample statistics for all samples. What "large enough" means can vary, but, as a general rule, a sample size of 30 will be sufficient for most cases.

The importance of the CLT cannot be overstated. The CLT enables you to apply your knowledge of the normal distribution (Chapter 5) when analyzing the variability in the sample means. That, in turn, allows you to make statistical judgments such as the ones discussed later in this chapter.

Figure 6.1 on page 124 illustrates that the CLT applies to all types of populations, regardless of their shape. In the figure, the effects of increasing sample size are shown for:

- A normally distributed population (left column)
- A uniformly distributed population in which the values are evenly distributed between the smallest and largest values (middle column)
- An exponentially distributed population in which the values are heavily skewed to the right (right column)

For the normally distributed population, the sampling distribution of the mean is always normally distributed, too. As the sample size increases, the variability of the sample means decreases, resulting in a narrowing of the graph.

For the other two populations, a **central limiting** effect causes the sample means to become more similar and the shape of the graphs to become more like a normal distribution. This effect happens initially more slowly for the heavily skewed exponential distribution than for the uniform distribution, but when the sample size is increased to 30, the sampling distributions of these two populations converge to the shape of the sampling distribution of the normal population.

Figure 6.1 illustrates the following conclusions about the sampling distribution of the mean:

- For most population distributions, regardless of shape, the sampling distribution of the mean is approximately normally distributed if samples of at least 30 observations are selected.
- If the population distribution is fairly symmetrical, the sampling distribution of the mean is approximately normally distributed if samples of at least 15 observations are selected.
- If the population is normally distributed, the sampling distribution of the mean is normally distributed, regardless of the sample size.

FIGURE 6.1

Three
populations

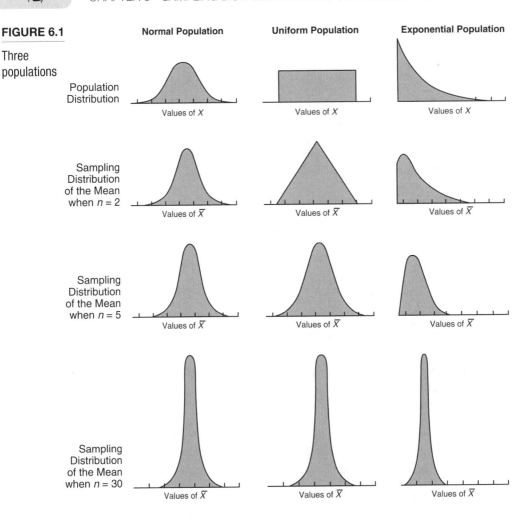

Sampling Distribution of the Proportion

CONCEPT   The probability distribution of a proportion for all possible
samples of a given size *n* for a given population.

INTERPRETATION   Section 5.2 explains that you can use a binomial distribution to determine probabilities for categorical variables that have only two
categories, traditionally labeled "success" and "failure." As the sample size
increases for such variables, you can use the normal distribution to approximate
the sampling distribution of the number of successes or the proportion of
successes.

As a general rule, you can use the normal distribution to approximate the
binomial distribution when the number of successes and the number of failures

are each at least five. For most cases in which you are estimating the proportion, the sample size is more than sufficient to meet the conditions for using the normal approximation.

6.2 Sampling Error and Confidence Intervals

By taking one sample and computing the results of a sample statistic, such as the mean, you create a *point estimate* of the population parameter. This single estimate will almost certainly be different if another sample is selected, as WORKED-OUT PROBLEM 1 illustrates.

WORKED-OUT PROBLEM 1   From a population of N = 200 order-filling times, you want to select 20 samples of n = 15 and calculate sample statistics. The population has a population mean = 69.637 and a population standard deviation = 10.411. One such set of 20 samples of n = 15 appears in the following table.

Order Time Population

Sample	Mean	Standard Deviation	Minimum	Median	Maximum	Range
1	66.12	9.21	47.20	65.00	87.00	39.80
2	73.30	12.48	52.40	71.10	101.10	48.70
3	68.67	10.78	54.00	69.10	85.40	31.40
4	69.95	10.57	54.50	68.00	87.80	33.30
5	73.27	13.56	54.40	71.80	101.10	46.70
6	69.27	10.04	50.10	70.30	85.70	35.60
7	66.75	9.38	52.40	67.30	82.60	30.20
8	68.72	7.62	54.50	68.80	81.50	27.00
9	72.42	9.97	50.10	71.90	88.90	38.80
10	69.25	10.68	51.10	66.50	85.40	34.30
11	72.56	10.60	60.20	69.10	101.10	40.90
12	69.48	11.67	49.10	69.40	97.70	48.60
13	64.65	9.71	47.10	64.10	78.50	31.40
14	68.85	14.42	46.80	69.40	88.10	41.30
15	67.91	8.34	52.40	69.40	79.60	27.20
16	66.22	10.18	51.00	66.40	85.40	34.40
17	68.17	8.18	54.20	66.50	86.10	31.90
18	68.73	8.50	57.70	66.10	84.40	26.70
19	68.57	11.08	47.10	70.40	82.60	35.50
20	75.80	12.49	56.70	77.10	101.10	44.40

From these results, you observe:

- The sample statistics differ from sample to sample. The sample means vary from 64.65 to 75.80, the sample standard deviations vary from 7.62 to 14.42, the sample medians vary from 64.10 to 77.10, and the sample ranges vary from 26.70 to 48.70.
- Some of the sample means are higher than the population mean of 69.637, and some of the sample means are lower than the population mean.
- Some of the sample standard deviations are higher than the population standard deviation of 10.411, and some of the sample standard deviations are lower than the population standard deviation.
- The variation in the sample range from sample to sample is much greater than the variation in the sample standard deviation.

Sample statistics almost always vary from sample to sample. This expected variation is called the sampling error.

Sampling Error

CONCEPT   The variation that occurs due to selecting a single sample from the population.

EXAMPLE   In polls, the plus-or-minus margin of the results, as in "42%, plus or minus 3%, said they were likely to vote for the incumbent."

INTERPRETATION   The size of the sampling error is primarily based on the variation in the population itself and on the size of the sample selected. Larger samples have less sampling error but are more costly to take.

In practice, only one sample is used as the basis for estimating a population parameter. To account for the differences in the results from sample to sample, you use a confidence interval estimate.

Confidence Interval Estimate

CONCEPT   A range with explicit lower and upper limits, stated with a specific degree of certainty, that represents an estimate of a population parameter.

INTERPRETATION   As WORKED-OUT PROBLEM 1 illustrates, sample statistics vary and can be less than or greater than the actual population parameter. A confidence interval estimate creates a range centered on a sample statistic and identifies the likelihood that this range includes the actual population parameter. To calculate this range, you need the calculated sample statistic being used to estimate the population parameter and the sampling distribution associated with the sample statistic.

important point

You must state the given degree of certainty, or **confidence**, when reporting an interval estimate—for example, an "interval estimate with 95% confidence" (which is sometimes phrased as a "95% confidence interval estimate"). An "interval estimate with 95% confidence" means that if all possible samples of the sample size that your single sample uses were taken, 95% percent of interval estimates calculated from those samples would include the population parameter, and 5% would not.

A properly stated degree of certainty answers the reservations some have when they wonder how you can reach conclusions about an entire population using only one sample. Any interval estimate stated without the degree of certainty is worthless.

Because you are estimating an interval using one sample and not precisely determining a value, you can never be 100% certain that your interval correctly estimates the population parameter. You trade off the level of confidence and the width of the interval: the greater the confidence that your interval will be correct, the wider the interval will be. Most use 95% confidence as an acceptable trade-off, although 99% confidence (resulting in a wider width) and 90% confidence (resulting in a narrower width) are also used in certain fields or types of analyses.

WORKED-OUT PROBLEM 2   You want to develop 95% confidence interval estimates for the mean from 20 samples of size 15 for the order-filling population data that WORKED-OUT PROBLEM 1 uses. For this problem, the population mean, 69.637, and the population standard deviation, 10.411, are known. One such set of 20 samples of $n = 15$ appears in the table on page 128.

Because the population mean is known, you will be able to compare the confidence interval estimate for the mean developed from each sample to the actual value of the population mean. From the results, you conclude:

- For sample 1, the sample mean is 66.12, the sample standard deviation is 9.21, and the interval estimate for the population mean is 60.85 to 71.39. This enables you to conclude with 95% certainty that the population mean is between 60.85 and 71.39. This is a correct estimate because the population mean of 69.637 is included within this interval.

- Although their sample means and standard deviations differ, the confidence interval estimates for samples 2 through 19 include the population mean of 69.637.

- For sample 20, the sample mean is 75.80, the sample standard deviation is 12.49, and the interval estimate for the population mean is 70.53 to 81.07 (highlighted in the results). This is an incorrect estimate because the population mean of 69.637 is not included within this interval.

Order Time Samples

Sample	Mean	Standard Deviation	Lower Limit	Upper Limit
1	66.12	9.21	60.85	71.39
2	73.30	12.48	68.03	78.57
3	68.67	10.78	63.40	73.94
4	69.95	10.57	64.68	75.22
5	73.27	13.56	68.00	78.54
6	69.27	10.04	64.00	74.54
7	66.75	9.38	61.48	72.02
8	68.72	7.62	63.45	73.99
9	72.42	9.97	67.15	77.69
10	69.25	10.68	63.98	74.52
11	72.56	10.60	67.29	77.83
12	69.48	11.67	64.21	74.75
13	64.65	9.71	59.38	69.92
14	68.85	14.42	63.58	74.12
15	67.91	8.34	62.64	73.18
16	66.22	10.18	60.95	71.49
17	68.17	8.18	62.90	73.44
18	68.73	8.50	63.46	74.00
19	68.57	11.08	63.30	73.84
20	75.80	12.49	**70.53**	**81.07**

That the percentage of the samples which have a confidence interval that includes the mean (95%, 19 out of 20) matches the level of confidence chosen (also 95%) is a coincidence. Other sets of 20 samples could have a percentage of the samples higher or lower than 95%, 19 out of 20. However, in the long run, 95% of all samples *would* produce a confidence interval that includes the mean.

6.3 Confidence Interval Estimate for the Mean Using the *t* Distribution (σ Unknown)

Calculating the confidence interval estimate for a mean requires knowing not only the sample mean but the population standard deviation as well. However, in nearly all cases, this parameter is unknown. The *t* distribution (reference 1) enables you to use the sample standard deviation, which can always be determined, to calculate confidence interval estimates.

t Distribution

CONCEPT   The sampling distribution that allows you to develop a confidence interval estimate of the mean using the sample standard deviation.

INTERPRETATION   The *t* distribution assumes that the variable being studied comes from a normally distributed population—something that may or may not be known. As a practical matter, though, as long as the sample size is large enough and the population is not very skewed, the *t* distribution can be used to estimate the population mean. However, with a small sample size or a skewed population distribution, you should verify that the variable does not violate the assumption of **normality**, that the variable being studied *is* normally distributed. (You can use a histogram, a boxplot, or a normal probability plot to see if the assumption is violated.)

Super Bowl Ads

WORKED-OUT PROBLEM 3   To further analyze the mean ad ratings for the 2021 Super Bowl broadcast, first used by Problem 4 in Chapter 2, you want to calculate the confidence interval estimates for the population mean ad rating for the first half and for the second half ad ratings.

Figure 6.2 presents the spreadsheet results for the confidence interval estimate for both the first half and second half mean ad rating:

FIGURE 6.2

Confidence interval estimates for the first half and second half mean Super Bowl ad ratings

	A	B
1	Confidence Interval Estimate	
2	for the First Half Mean Ad Rating	
3	Data	
4	Sample Standard Deviation	0.7020
5	Sample Mean	5.7893
6	Sample Size	28
7	Confidence Level	95%
8		
9	Intermediate Calculations	
10	Standard Error of the Mean	0.1327
11	Degrees of Freedom	27
12	t Value	2.0518
13	Interval Half Width	0.2722
14		
15	Confidence Interval	
16	Interval Lower Limit	5.5171
17	Interval Upper Limit	6.0615

	A	B
1	Confidence Interval Estimate	
2	for the Second Half Mean Ad Rating	
3	Data	
4	Sample Standard Deviation	0.7093
5	Sample Mean	5.5345
6	Sample Size	29
7	Confidence Level	95%
8		
9	Intermediate Calculations	
10	Standard Error of the Mean	0.1317
11	Degrees of Freedom	28
12	t Value	2.0484
13	Interval Half Width	0.2698
14		
15	Confidence Interval	
16	Interval Lower Limit	5.2647
17	Interval Upper Limit	5.8043

To evaluate the assumption of normality necessary to use these estimates, you use the boxplots created to solve WORKED-OUT PROBLEM 15 in Chapter 3. These boxplots have slight right-skewness because the tail on the right is longer than the one on the left. However, given the relatively large sample size, you conclude that any departure from the normality assumption will not seriously affect the validity of the confidence interval estimate.

Therefore, based on these results that use 95% confidence, you conclude that the mean ad rating for the first half is between 5.52 and 6.06 and that the mean ad rating for the second half is between 5.26 and 5.80.

equation blackboard (optional)

interested in math?

You use the symbols $\bar{X}$ (sample mean), μ (population mean), S (sample standard deviation), and n (sample size), introduced earlier, and the new symbols α (alpha) and t_{n-1}, which represents the critical value of the t distribution with $n-1$ degrees of freedom for an area of $\alpha/2$ in the upper tail, to express the confidence interval for the mean as a formula, in cases in which the population standard deviation, σ, is unknown. The symbol alpha, α, is equivalent to 1 minus the confidence percentage. For 95% confidence, α is 0.05 $(1 - 0.95)$, and $\alpha/2$, the upper tail area, is 0.025.

Using these symbols, you create the following equation:

$$\bar{X} \pm t_{n-1} \frac{S}{\sqrt{n}}$$

or, expressed as a range:

$$\bar{X} - t_{n-1} \frac{S}{\sqrt{n}} \leq \mu \leq \bar{X} + t_{n-1} \frac{S}{\sqrt{n}}$$

For WORKED-OUT PROBLEM 3, for the first half ad ratings, $\bar{X} = 5.7893$ and $S = 0.7020$, with 27 degrees of freedom (the sample size, 28, minus 1). Using 95% confidence, α is 0.05, and the area in the upper tail of the t distribution is 0.025 (0.05/2). Using Table C.2 in Appendix C, the critical value for the row with 27 degrees of freedom and the column with an area of 0.025 is 2.0518. Substituting these values yields the following result:

$$\bar{X} \pm t_{n-1} \frac{S}{\sqrt{n}} = 5.7893 \pm (2.0518) \frac{0.7020}{\sqrt{28}}$$

$$= 5.7893 \pm 0.2722$$

$$5.52 \leq \mu \leq 6.06$$

The confidence interval for the first half ad ratings mean is estimated to be between 5.52 and 6.06 with 95% confidence.

spreadsheet solution

Confidence Interval Estimate for the Mean When σ Is Unknown

Chapter 6 Sigma Unknown contains the two spreadsheets that WORKED-OUT PROBLEM 3 uses to calculate a confidence interval estimate for the population mean when the population standard deviation is unknown. The two spreadsheets (shown in Figure 6.2) are identical, other than data values entered in cells B4 through B7.

Best Practices

Use the **T.INV.2T(1 – *confidence level, degrees of freedom*)** function to calculate the critical value of the *t* distribution. Then use that critical value to calculate the confidence interval for the mean when σ is unknown. (For a confidence level of 95%, enter 0.05 as the value for **1 – *confidence level*.**)

How-Tos

For other problems, change the sample standard deviation, sample mean, sample size, and confidence level in cells B4 through B7 in one of the spreadsheets.

Advanced Technique ADV4 in Appendix E explains how to modify a **Chapter 6 Sigma Unknown** spreadsheet for use with unsummarized data, such as the data for problems 1 through 3 at the end of this chapter.

6.4 Confidence Interval Estimation for Categorical Variables

For a categorical variable, you can develop a confidence interval to estimate the proportion of successes in a given category.

Confidence Interval Estimation for the Proportion

CONCEPT   The sampling distribution of the proportion that allows you to develop a confidence interval estimate of the proportion using the sample proportion of successes, *p*. The sample statistic *p* follows a binomial distribution that can be approximated by the normal distribution for most studies.

EXAMPLE   The proportion of voters who would vote for a certain candidate in an election, the proportion of consumers who own a particular brand of smartphone, the proportion of medical tests in a hospital that need to be repeated.

INTERPRETATION   This type of confidence interval estimate uses the sample proportion of successes, p, equal to the number of successes divided by the sample size, to estimate the population proportion. (Categorical variables have no population means.)

For a given sample size, confidence intervals for proportions are wider than those for numerical variables. With continuous variables, the measurement on each respondent contributes more information than for a categorical variable. In other words, a categorical variable with only two possible values is a very crude measure compared with a continuous variable, so each observation contributes only a little information about the parameter being estimated.

WORKED-OUT PROBLEM 4   The owner of a restaurant that features Continental cuisine seeks to learn more about the proportion of guests who, during the Friday-to-Sunday weekend time, order beef entrées. From a sample of 100 customers in which 36 customers ordered beef entrées, you want to calculate the confidence interval estimates for the population proportion of customers who order beef entrées.

Figure 6.3 presents the spreadsheet results for the confidence interval estimate of the population proportion, using 95% confidence. Based on these results, you estimate that between 26.59% and 45.41% of Friday-to-Sunday weekend time customers order beef entrées.

FIGURE 6.3

Confidence interval estimates for the proportion of customers who order beef entrées

	A	B
1	Confidence Interval Estimate for the Proportion	
2		
3	Data	
4	Sample Size	100
5	Number of Successes	36
6	Confidence Level	95%
7		
8	Intermediate Calculations	
9	Sample Proportion	0.3600
10	Z Value	-1.9600
11	Standard Error of the Proportion	0.0480
12	Interval Half Width	0.0941
13		
14	Confidence Interval	
15	Interval Lower Limit	0.2659
16	Interval Upper Limit	0.4541

interested in math?

You use the symbols p (sample proportion of success), n (sample size), and Z (Z score), previously introduced, and the symbol π for the population proportion, to assemble the equation for the confidence interval estimate for the proportion:

$$p \pm Z\sqrt{\frac{p(1-p)}{n}}$$

Or, expressed as a range:

$$p - Z\sqrt{\frac{p(1-p)}{n}} \leq \pi \leq p + Z\sqrt{\frac{p(1-p)}{n}}$$

Z = Critical value from the normal distribution

For WORKED-OUT PROBLEM 4, $n = 100$ and $p = 36/100 = 0.36$. For a 95% level of confidence, the lower tail area of 0.025 provides a Z value from the normal distribution of -1.96, and the upper tail area of 0.025 provides a Z value from the normal distribution of $+1.96$.

Substituting these numbers into the preceding equation yields the following results:

$$p \pm Z\sqrt{\frac{p(1-p)}{n}} = 0.36 \pm (1.96)\sqrt{\frac{(0.36)(0.64)}{100}}$$

$$= 0.36 \pm (1.96)(0.0480)$$

$$= 0.36 \pm 0.0941$$

$$0.2659 \leq \pi \leq 0.4541$$

Based on these results, you estimate that between 26.59% and 45.41% of Friday-to-Sunday weekend time customers order beef entrées.

spreadsheet solution

Confidence Interval Estimate for the Proportion

Chapter 6 Proportion contains the spreadsheet that WORKED-OUT PROBLEM 4 uses to calculate a confidence interval estimate for the population proportion (see Figure 6.3).

Best Practices

Use the **NORM.S.INV((1–*confidence level*)/2)** function to calculate the critical value of the normal distribution. Then, use that critical value to calculate the confidence interval for the proportion. (For a confidence level of 95%, enter 0.05 as the value for **1–***confidence level*.)

Use the absolute value **ABS(*value or formula*)** function to help calculate the absolute value of the half-width of the confidence interval.

How-Tos

For other problems, change the sample size, number of successes, and confidence level in cells B4 through B6.

6.5 Confidence Interval Estimation When Normality Cannot Be Assumed

The confidence interval estimation methods for population parameters that Sections 6.3 and 6.4 discuss assume that the population being analyzed is normally distributed. When you cannot assume normality, you can use bootstrap estimation to estimate a population parameter.

Bootstrapping

CONCEPT   Using sampling with replacement (see Section 1.6) to create many samples ("resamples") from an initial sample created without replacement.

INTERPRETATION   Bootstrapping are techniques to create a simulated population with characteristics similar to a normal population. Using bootstrapping eliminates the need to make assumptions about population data. For data analysis methods that assume a normal population, bootstrapping can often be used to create a simulated population from a set of data that cannot be assumed to follow a normal distribution.

Bootstrap Estimation

CONCEPT   The use of bootstrapping to create a data set that can be used to estimate a population parameter.

EXAMPLE   Bootstrap estimation of the population mean might consist of these steps:

1. Select a random sample of size n *without replacement* from a population of size N.
2. Resample the initial sample by selecting n values *with replacement* from the n values in the initial sample and compute the sample means for this resample.
3. Repeat step 2 m number of times to produce m resamples.
4. Construct the resampling distribution of the sample mean from each of the m samples.
5. Construct an ordered array of the entire set of resampled means.
6. Find the values that exclude the smallest $\alpha/2 \times 100\%$ of means and the largest $\alpha/2 \times 100\%$ of means. These values become the lower and upper limits of the bootstrap confidence interval estimate of the population mean with $(1 - \alpha)\%$ confidence.

INTERPRETATION   When using bootstrap estimation, you typically use software to select a very large number of resamples and to perform the bootstrap estimation. Using a very large number of resamples reduces the sampling error but increases the processing time. What "very large" means partially depends on the population size and may be limited by computing resources available for the analysis.

WORKED-OUT PROBLEM 5   An insurance company seeks to reduce the amount of time it takes to approve life insurance applications. You collect data by selecting a random sample of 27 approved policies during a period of one month. This sample contains the following total processing times (in days):

| 73 | 19 | 16 | 64 | 28 | 28 | 31 | 90 | 60 | 56 | 31 | 56 | 22 | 18 |
| 45 | 48 | 17 | 17 | 17 | 91 | 92 | 63 | 50 | 51 | 69 | 16 | 17 |

Insurance

A 90% confidence interval estimate for the population mean processing time indicates that the mean processing time for the population of life insurance applications is between 35.59 and 52.19 days. This estimate assumes that the population of processing times is normally distributed, but a boxplot and a normal probability plot (not shown) indicate that the population is right-skewed. This raises questions about the validity of the confidence interval, and you decide to use a bootstrap estimation method that will use 100 resamples.

The following list shows the first resample that is based on the initial sample of 27 processing times:

16	16	16	17	17	17	17	17	19	22	28	31	31	48
51	56	56	60	60	64	64	64	69	**73**	**73**	90	92	

This first resample omits some values (18, 45, 50, 63, and 91) that appear in the initial sample and repeats some other values more times than they appear in the initial sample. For example, the value 73 (highlighted) appears only once in the initial sample but twice in the resample as a result of resampling *with replacement*. After taking 100 resamples of the processing times, you compute sample means and construct the following ordered array of the sample means for the 100 resamples:

31.5926	33.9259	35.4074	36.5185	**36.6296**	36.9630	37.0370	37.0741
37.1481	37.3704	37.9259	38.1111	38.1481	38.2222	38.2963	38.7407
38.8148	38.8519	38.8889	39.0000	39.1852	39.3333	39.3704	39.6667
40.1481	40.5185	40.6296	40.9259	40.9630	41.2593	41.2963	41.7037
41.8889	42.0741	42.1111	42.1852	42.8519	43.0741	43.1852	43.3704
43.4444	43.7037	43.8148	43.8519	43.8519	43.9259	43.9630	44.1481
44.4074	44.5556	44.7778	45.0000	45.4444	45.5185	45.5556	45.6667
45.7407	45.8519	45.9630	45.9630	46.0000	46.1111	46.2963	46.2963
46.3333	46.3333	46.4815	46.6667	46.7407	46.9630	47.0741	47.2222
47.2963	47.3704	47.4815	47.4815	47.5556	47.6667	47.8519	48.5185
48.8889	49.0000	49.2222	49.4444	49.4815	49.4815	49.6296	49.6296
49.7407	50.2963	50.4074	50.5926	50.9259	51.4074	51.4815	**51.5926**
51.9259	52.3704	53.4074	54.3333				

You need to identify the fifth-smallest and fifth-largest values to exclude the smallest and largest 5% of the resample means (90% confidence is equal to $\alpha = 0.1$; $\alpha/2 = 0.05$, and $0.05 \times 100\% = 5$). In the ordered array, the fifth-smallest value is 36.6296 and the fifth-largest value is 51.5926 (highlighted). Therefore, the 90% bootstrap confidence interval estimate of the population mean processing time is 36.6296 to 51.5926 days.

Important Equations

Confidence Interval for the Mean with σ Unknown:

$$\bar{X} \pm t_{n-1} \frac{S}{\sqrt{n}}$$

or

$$\bar{X} \pm t_{n-1} \frac{S}{\sqrt{n}} \leq \mu \leq \bar{X} + t_{n-1} \frac{S}{\sqrt{n}}$$

Confidence Interval Estimate for the Proportion:

$$p \pm Z \sqrt{\frac{p(1-p)}{n}}$$

or

$$p \pm Z \sqrt{\frac{p(1-p)}{n}} \leq \pi \leq p + Z \sqrt{\frac{p(1-p)}{n}}$$

One-Minute Summary

Which confidence interval estimate you use depends on the type of variable being studied:

- Use the confidence interval estimate for the mean for a numerical variable.
- Use the confidence interval estimate for the proportion for a categorical variable.

Test Yourself
Short Answers

1. The sampling distribution of the mean can be approximated by the normal distribution:
    a. as the number of samples gets "large enough"
    b. . as the sample size (number of observations in each sample) gets large enough
    c. as the size of the population standard deviation increases
    d. as the size of the sample standard deviation decreases

2. The sampling distribution of the mean requires _____ sample size to reach a normal distribution if the population is skewed than if the population is symmetrical.

   a.   the same

   b.   a smaller

   c.   a larger

   d.   The two distributions cannot be compared.

3. Which of the following is true regarding the sampling distribution of the mean for a large sample size?

   a.   It has the same shape and mean as the population.

   b.   It has a normal distribution with the same mean as the population.

   c.   It has a normal distribution with a different mean from the population.

4. For samples of $n = 30$, for most populations, the sampling distribution of the mean will be approximately normally distributed:

   a.   regardless of the shape of the population

   b.   if the shape of the population is symmetrical

   c.   if the standard deviation of the mean is known

   d.   if the population is normally distributed

5. For samples of $n = 1$, the sampling distribution of the mean will be normally distributed:

   a.   regardless of the shape of the population

   b.   if the shape of the population is symmetrical

   c.   if the standard deviation of the mean is known

   d.   if the population is normally distributed

6. A 99% confidence interval estimate can be interpreted to mean that:

   a.   If all possible samples are taken and confidence interval estimates are developed, 99% of them would include the true population mean somewhere within their interval.

   b.   You have 99% confidence that you have selected a sample whose interval does include the population mean.

   c.   Both a and b are true.

   d.   Neither a nor b is true.

7. Which of the following statements is false?

   a.   There is a different critical value for each level of alpha (α).

   b.   Alpha (α) is the proportion in the tails of the distribution that is outside the confidence interval.

c. You can construct a 100% confidence interval estimate of μ.

d. In practice, the population mean is the unknown quantity that is to be estimated.

8. Sampling distributions describe the distribution of:

   a. parameters
   b. statistics
   c. both parameters and statistics
   d. neither parameters nor statistics

9. In the construction of confidence intervals, if all other quantities are unchanged, an increase in the sample size will lead to _____ interval.

   a. a narrower
   b. a wider
   c. a less significant
   d. the same

10. As an aid to the establishment of personnel requirements, the manager of a bank wants to estimate the mean number of people who arrive at the bank during the two-hour lunch period from 12 noon to 2 p.m. The director randomly selects 64 different two-hour lunch periods from 12 noon to 2 p.m. and determines the number of people who arrive for each. For this sample, $\bar{X}$ = 49.8 and S = 5. Which of the following assumptions is necessary in order for a confidence interval to be valid?

   a. The population sampled from has an approximate normal distribution.
   b. The population sampled from has an approximate t distribution.
   c. The mean of the sample equals the mean of the population.
   d. None of these assumptions are necessary.

11. A university dean is interested in determining the proportion of students who are planning to attend graduate school. Rather than examine the records for all students, the dean randomly selects 200 students and finds that 118 of them are planning to attend graduate school. The 95% confidence interval for p is 0.59 ± 0.07. Interpret this interval.

   a. You are 95% confident that the true proportion of all students planning to attend graduate school is between 0.52 and 0.66.
   b. There is a 95% chance of selecting a sample that finds that between 52% and 66% of the students are planning to attend graduate school.
   c. You are 95% confident that between 52% and 66% of the sampled students are planning to attend graduate school.
   d. You are 95% confident that 59% of the students are planning to attend graduate school.

12. In estimating the population mean with the population standard deviation unknown, if the sample size is 12, there will be _____ degrees of freedom.

13. The Central Limit Theorem is important in statistics because

    a.   it states that the population will always be approximately normally distributed.

    b.   it states that the sampling distribution of the sample mean is approximately normally distributed for a large sample size n, regardless of the shape of the population.

    c.   it states that the sampling distribution of the sample mean is approximately normally distributed for any population, regardless of the sample size.

    d.   for a sample of any size, it says the sampling distribution of the sample mean is approximately normal.

14. For samples of $n = 15$, the sampling distribution of the mean will be normally distributed:

    a.   regardless of the shape of the population

    b.   if the shape of the population is symmetrical

    c.   if the standard deviation of the mean is known

    d.   if the population is normally distributed

Answer True or False:

15. Other things being equal, as the confidence level for a confidence interval increases, the width of the interval increases.

16. As the sample size increases, the effect of an extreme value on the sample mean becomes smaller.

17. A sampling distribution is defined as the probability distribution of possible sample sizes that can be observed from a given population.

18. The t distribution is used to construct confidence intervals for the population mean when the population standard deviation is unknown.

19. In the construction of confidence intervals, if all other quantities are unchanged, an increase in the sample size will lead to a wider interval.

20. The confidence interval estimate that is constructed will always correctly estimate the population parameter.

Answers to Test Yourself Short Answers

1. b   6. c

2. c   7. c

3. b   8. b

4. a   9. a

5. d   10. d

11. a	16. True
12. 11	17. False
13. b	18. True
14. b	19. False
15. True	20. False

Problems

MPG

1. The file **MPG** contains the overall miles per gallon for 2021 sedans. The data are

Sedan	Overall MPG	Sedan	Overall MPG
Alfa Romeo Giulia	27	Lexus ES	25
Audi A6	26	Mazda 6	28
Audi A8	21	Mercedes-Benz C Class	26
BMW 5 Series	26		
BMW X1	26	Mitsubishi Mirage	37
Chevrolet Malibu	29	Nissan Altima	31
Chrysler 300	22	Nissan Versa	32
Genesis 370	26	Subaru Impreza	30
Honda Accord	31	Subaru Legacy	28
Hyundai Accent	33	Toyota Avalon	42
Infiniti Q50	22	Toyota Camry	32
Kia Cadenza	24	Toyota Corolla	33
Kia Rio	33	Volkswagen Jetta	34
Land Rover Range Rover	17	Volvo S90	23

Source: Data extracted from "Cars," https://consumerreprts.org/cars.

Construct a 95% confidence interval estimate of the mean overall MPG for 2021 sedans.

Protein

2. The following table lists the calories, protein, percentage of calories from fat, percentage of calories from saturated fat, and cholesterol of popular protein foods (fresh red meats, poultry, and fish).

Food	Calories	Protein	Pctage Calories from Fat	Pctage Calories from Sat. Fat	Cholesterol
Beef, ground, extra lean	250	25	58	23	82
Beef, ground, regular	287	23	66	26	87
Beef, round	184	28	24	12	82
Brisket	263	28	54	21	91
Flank steak	244	28	51	22	71
Lamb leg roast	191	28	38	16	89
Lamb loin chop, broiled	215	30	42	17	94
Liver, fried	217	27	36	12	482
Pork loin roast	240	27	52	18	90
Sirloin	208	30	37	15	89
Spareribs	397	29	67	27	121
Veal cutlet, fried	183	33	42	20	127
Veal rib roast	175	26	37	15	131
Chicken, with skin, roasted	239	27	51	14	88
Chicken, no skin, roast	190	29	37	10	89
Turkey, light meat, no skin	157	30	18	6	69
Clams	98	16	6	0	39
Cod	98	22	8	1	74
Flounder	99	21	12	2	54
Mackerel	199	27	77	20	100
Ocean perch	110	23	13	3	53
Salmon	182	27	24	5	93
Scallops	112	23	8	1	56
Shrimp	116	24	15	2	156
Tuna	181	32	41	10	48

Source: U.S. Department of Agriculture.

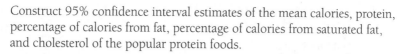

Construct 95% confidence interval estimates of the mean calories, protein, percentage of calories from fat, percentage of calories from saturated fat, and cholesterol of the popular protein foods.

Domestic Beer

3. The **Domestic Beer** file contains the percentage alcohol, number of calories per 12 ounces, and number of carbohydrates (in grams) per 12 ounces for 157 of the best-selling domestic beers in the United States. (Data extracted from "Find Out How Many Calories in Beer?" https://beer100.com/beer-calories.)

   a. Construct 95% confidence interval estimates of the mean percentage alcohol, mean number of calories per 12 ounces, and mean number of carbohydrates (in grams) per 12 ounces.
   b. Do you need to assume that the variables in (a) are normally distributed in order to construct the confidence intervals in (a)?

4. A Pew Research Center survey studied the key issues for employed adults who have been working at home some or all of the time. Of 101 respondents, 36 said that they found working at home difficult. Construct a 95% confidence interval estimate of the proportion of employed adults who found working at home difficult.

5. Churning, the loss of customers to a competitor, is a problem for all companies, especially telecommunications companies. Market researchers for a telecommunications company collected data from a sample of 5,517 customers of the company and discovered that 1,756 customers churned in the past month. Construct a 95% confidence interval estimate of the proportion of customers who churned in the past month.

6. A national pizza chain seeks to expand and open a store in the small town that is home to a regional public university whose students are well known for being customers of the local independent pizza restaurant. As part of its planning, the pizza chain conducts a survey of students to determine pizza preferences. The chain selects a sample of 220 students attending the university. The results indicated that 145 of the 220 students preferred the local pizza restaurant to the national chain. Construct a 95% confidence interval estimate of the proportion of students who preferred the local pizza restaurant to the national chain.

Answers to Problems

1. $26.17 \leq \mu \leq 30.43$.
2. Calories: $164.80 \leq \mu \leq 222.00$; protein: $24.97 \leq \mu \leq 28.07\%$; calories from fat: $28.23 \leq \mu \leq 44.89\%$; calories from saturated fat: $9.25 \leq \mu \leq 16.19$; cholesterol: $67.68 \leq \mu \leq 136.72$.
3. a. Alcohol percentage: $5.06 \leq \mu \leq 5.48$; calories: $148.75 \leq \mu \leq 162.56$; carbohydrates: $11.27 \leq \mu \leq 12.84$.

b.   The sample size is large ($n = 157$), so the use of the t distribution to construct the confidence interval is appropriate because the validity will not be affected.

4. $0.2630 \leq \pi \leq 0.4498$.

5. $0.3060 \leq \pi \leq 0.3306$.

6. $0.5965 \leq \pi \leq 0.7217$

References

1. Berenson, M. L., D. M. Levine, and K. A. Szabat, and D. Stephan. *Basic Business Statistics: Concepts and Applications*, 15th edition. Hoboken, NJ: Pearson Education, 2023.

2. Cochran, W. G. *Sampling Techniques*, 3rd edition. New York: Wiley, 1977.

3. Diaconis, P. and B. Efron. "Computer-Intensive Methods in Statistics," *Scientific American*, 248, 1983, pp. 116–130.

4. Efron, B, and R. Tibshirani. *An Introduction to the Bootstrap*. Boca Raton, FL: Chapman and Hall/CRC, 1995.

5. Gunter, B. "Bootstrapping: How to Make Something from Almost Nothing and Get Statistically Valid Answers Part I: Brave New World," *Quality Progress*, 24, December 1991, pp. 97–103.

6. Levine, D. M., D. Stephan, and K. A. Szabat. *Statistics for Managers Using Microsoft Excel*, 9th edition. Hoboken, NJ: Pearson Education, 2021.

7. Levine, D. M., P. P. Ramsey, and R. K. Smidt. *Applied Statistics for Engineers and Scientists Using Microsoft Excel and Minitab*. Upper Saddle River, NJ: Prentice Hall, 2001.

8. Varian, H. "Bootstrap Tutorial," *Mathematica Journal*, 9, 2005, pp. 768–775.

Science progresses by first stating tentative explanations, or hypotheses, about natural phenomena and then by proving (or disproving) those hypotheses through investigation and testing. Statisticians have adapted this scientific method by developing an inferential method called **hypothesis testing** that evaluates a claim made about the value of a population parameter by using a sample statistic. In this chapter, you learn the basic concepts and principles of hypothesis testing and the statistical assumptions necessary for performing hypothesis testing.

7.1 The Null and Alternative Hypotheses

Unlike the broader hypothesis testing of science, statistical hypothesis testing always involves evaluating a claim made about the value of a population parameter. This claim is stated as a pair of statements: the null hypothesis and the alternative hypothesis.

Null Hypothesis

CONCEPT   The statement that a population parameter is equal to a specific value or that the population parameters from two or more groups are equal.

EXAMPLES   "The population mean time to answer customer emails was 4 hours last year," "the mean height for women is the same as the mean height for men," "at a restaurant, the proportion of orders filled correctly for drive-through customers is the same as the proportion of orders filled correctly for sit-down customers."

important point

INTERPRETATION   The null hypothesis always expresses an equality and is always paired with another statement, the alternative hypothesis. A null hypothesis is considered true until evidence indicates otherwise. If you can conclude that the null hypothesis is false, then the alternative hypothesis must be true.

You use the symbol H_0 to identify the null hypothesis and write a null hypothesis using an equal sign and the symbol for the population parameter, as in H_0: $\mu = 4$ or H_0: $\mu_1 = \mu_2$ or H_0: $\pi_1 = \pi_2$. (Remember that in statistics, the symbol π represents the population proportion and not the geometric ratio of the circumference to the diameter of a circle.)

Alternative Hypothesis

CONCEPT   The statement paired with a null hypothesis that is mutually exclusive to the null hypothesis.

EXAMPLES   "The population mean for the time to answer customer emails was not 4 hours last year" (which would be paired with the example for the null hypothesis in the preceding section); "the mean height for women is not the same as the mean height for men" (paired with the second example for the null hypothesis); "at a restaurant, the proportion of food orders filled correctly for drive-through customers is not the same as the proportion of food orders filled correctly for sit-down customers" (paired with the third example for the null hypothesis).

INTERPRETATION   The alternative hypothesis is typically the idea you are studying concerning your data. The alternative hypothesis always expresses an inequality, either between a population parameter and a specific value or between two or more population parameters and is always paired with the null hypothesis. You use the symbol H_1 to identify the alternative hypothesis and write an alternative hypothesis using either a not-equal sign or a less than or greater than sign, along with the symbol for the population parameter, as in H_1: $\mu \neq 4$ or H_1: $\mu_1 \neq \mu_2$ or H_0: $\pi_1 \neq \pi_2$.

The alternative hypothesis represents the conclusion reached by rejecting the null hypothesis. You reject the null hypothesis if evidence from the sample statistic indicates that the null hypothesis is unlikely to be true. However, if you cannot reject the null hypothesis, you cannot claim to have proven the null hypothesis. Failure to reject the null hypothesis means (only) that you have failed to prove the alternative hypothesis.

7.2 Hypothesis Testing Issues

In hypothesis testing, you use the sample statistic to estimate the population parameter stated in the null hypothesis. For example, to evaluate the null hypothesis "the population mean time to answer customer emails was 4 hours last year," you would use the sample mean time to estimate the population mean time. As Chapter 6 establishes, a sample statistic is unlikely to be identical to its corresponding population parameter.

If the sample statistic is not the same as the population parameter, as it almost never is, the issue of whether to reject the null hypothesis involves deciding how different the sample statistic is from its corresponding population parameter. (In the case of two groups, the issue can be expressed, under certain conditions, as deciding how different the sample statistics of each group are from each other.)

Without a rigorous procedure that includes a clear definition of a difference, you would find it hard to decide on a consistent basis whether a null hypothesis is false and, therefore, whether to reject or not reject the null hypothesis. Statistical hypothesis testing methods provide such definitions and enable you to restate the decision-making process as the probability of computing a given sample statistic, if the null hypothesis were true, through the use of a test statistic and a risk factor.

Test Statistic

CONCEPT The value based on the sample statistic and the sampling distribution for the sample statistic.

EXAMPLES Test statistic for the difference between two sample means (Chapter 8), test statistic for the difference between two sample proportions (Chapter 8), test statistic for the difference between the means of more than two groups (Chapter 9), test statistic for the slope (Chapter 10).

INTERPRETATION If you are testing whether the mean of a population was equal to a specific value, the sample statistic is the sample mean. The test statistic is based on the difference between the sample mean and the value of the population mean stated in the null hypothesis. This test statistic follows the t distribution that Section 6.3 explains and Sections 8.2 and 8.3 apply.

If you are testing whether the mean of population 1 is equal to the mean of population 2, the sample statistic is the difference between the mean in sample 1 and the mean in sample 2. The test statistic is based on the difference between the mean in sample 1 and the mean in sample 2. Under certain circumstances, this test statistic also follows the *t* distribution.

Figure 7.1 shows that the sampling distribution of the test statistic is divided into a **region of rejection** (also known as the critical region) and a **region of nonrejection**. If the test statistic falls into the region of nonrejection, the null hypothesis is not rejected.

FIGURE 7.1

Parts of the sampling distribution of the test statistic

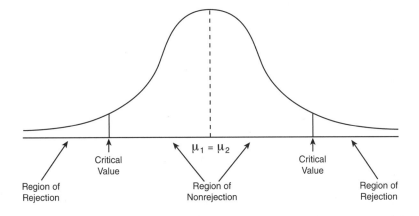

The region of rejection contains the values of the test statistic that are unlikely to occur if the null hypothesis is true. If the null hypothesis is false, these values are likely to occur. Therefore, if you observe a value of the test statistic that falls into the rejection region, you reject the null hypothesis because that value is unlikely if the null hypothesis is true.

To make a decision concerning the null hypothesis, you first determine the critical value of the test statistic that separates the nonrejection region from the rejection region. You determine the critical value by using the appropriate sampling distribution and deciding on the risk you are willing to take of rejecting the null hypothesis when it is true.

Practical Significance Versus Statistical Significance

important point

Another issue in hypothesis testing concerns the distinction between a statistically significant difference and a practical significant difference. Given a large enough sample size, it is always possible to detect a statistically significant difference. This is because no two things in nature are exactly equal. So, with a large enough sample size, you can always detect the natural difference between two populations. You need to be aware of the real-world practical implications of the statistical significance.

7.3 **Decision-Making Risks**

In hypothesis testing, you always face the possibility that either you will wrongly reject the null hypothesis or wrongly not reject the null hypothesis. These possibilities are called type I and type II errors, respectively.

Type I Error

CONCEPT   The error that occurs if the null hypothesis, H_0, is rejected when it is true and should not be rejected.

INTERPRETATION   The risk, or probability, of a type I error occurring is identified by the Greek lowercase alpha, α. Alpha is also known as the level of significance of the statistical test. Traditionally, you control the probability of a type I error by deciding the risk level α you are willing to tolerate of rejecting the null hypothesis when it is true.

Because you specify the level of significance before performing the hypothesis test, the risk of committing a type I error, α, is directly under your control. The most common α values are 0.01, 0.05, and 0.10, and researchers traditionally select a value of 0.05 or smaller.

When you specify the value for α, you determine the rejection region, and using the appropriate sampling distribution, the critical value or values that divide the rejection and nonrejection regions are determined.

Type II Error

CONCEPT   The error that occurs if the null hypothesis, H_0, is not rejected when it is false and should be rejected.

INTERPRETATION   The risk, or probability, of a type II error occurring is identified by the Greek lowercase beta, β. The probability of a type II error depends on the size of the difference between the value of the population parameter stated in the null hypothesis and the actual population value. Unlike a type I error, a type II error is not directly established by you. Because large differences are easier to find, as the difference between the value of the population parameter stated in the null hypothesis and its corresponding population parameter increases, the probability of a type II error decreases. Therefore, if the difference between the value of the population parameter stated in the null hypothesis and the corresponding parameter is small, the probability of a type II error will be large.

The arithmetic complement of beta, $1 - \beta$, is known as the **power of the test** and represents the probability of rejecting the null hypothesis when it is false and should be rejected.

Risk Trade-Off

Table 7.1 summarizes the types of errors and their associated risks. The probabilities of the two types of errors have an inverse relationship. When you decrease α, you always increase β, and when you decrease β, you always increase α.

TABLE 7.1
Risks and Decisions in Hypothesis Testing

		Actual Situation	
		H_0 true	H_0 false
Statistical Decision	**Do not reject H_0**	Correct decision Confidence = $1 - \alpha$	Type II error P(Type II error) = β
	Reject H_0	Type I error P(Type I error) = α	Correct decision Power = $1 - \beta$

One way you can lower β without affecting the value of α is to increase the sample size. Larger sample sizes generally permit you to detect even very small differences between the hypothesized and actual values of the population parameter. For a given level of α, increasing the sample size will decrease β and therefore increase the power of the test to detect that the null hypothesis, H_0, is false.

In establishing a value for α, you need to consider the negative consequences of a type I error. If these consequences are substantial, you can set $\alpha = 0.01$ instead of 0.05 and tolerate the greater β that results. If the negative consequences of a type II error most concern you, you can select a larger value for α (for example, 0.05 rather than 0.01) and benefit from the lower β that you will have.

7.4 Performing Hypothesis Testing

When you perform a hypothesis test, you should follow the steps of hypothesis testing in this order:

1. State the null hypothesis, H_0, and the alternative hypothesis, H_1.
2. Evaluate the risks of making type I and II errors and choose the level of significance, α, and the sample size as appropriate.
3. Determine the appropriate test statistic and sampling distribution to use and identify the critical values that divide the rejection and nonrejection regions.
4. Collect the data, calculate the appropriate test statistic, and determine whether the test statistic has fallen into the rejection region or the nonrejection region.
5. Make the proper statistical inference. Reject the null hypothesis if the test statistic falls into the rejection region. Do not reject the null hypothesis if the test statistic falls into the nonrejection region.

The *p*-Value Approach to Hypothesis Testing

Most modern statistical software, including the functions found in spreadsheet programs and calculators, can calculate the probability value known as the *p*-value that you can also use to determine whether to reject the null hypothesis.

p-Value

CONCEPT   The probability of computing a test statistic equal to or more extreme than the sample results, given that the null hypothesis, H_0, is true.

INTERPRETATION   The *p*-value is the smallest level at which H_0 can be rejected for a given set of data. You can consider the *p*-value the actual risk of having a type I error for a given set of data. Using *p*-values, the decision rules for rejecting the null hypothesis are as follows:

important point

- If the *p*-value is greater than or equal to α, do not reject the null hypothesis.
- If the *p*-value is less than α, reject the null hypothesis.
- Many people confuse this rule, mistakenly believing that a high *p*-value is reason for rejection. You can avoid this confusion by remembering the following saying:

"If the *p*-value is low, then H_0 must go."

In practice, most researchers today use *p*-values for several reasons, including efficiency of the presentation of results. The *p*-value is also known as the **observed level of significance**. When using *p*-values, you can restate the steps of hypothesis testing as follows:

1. State the null hypothesis, H_0, and the alternative hypothesis, H_1.
2. Evaluate the risks of making type I and II errors and choose the level of significance, α, and the sample size as appropriate.
3. Collect the data and calculate the sample value of the appropriate test statistic.
4. Calculate the *p*-value based on the test statistic and compare the *p*-value to α.
5. Make the proper statistical inference. Reject the null hypothesis if the *p*-value is less than α. Do not reject the null hypothesis if the *p*-value is greater than or equal to α.

7.5   Types of Hypothesis Tests

Your choice of which statistical test to use when performing hypothesis testing is influenced by the following factors:

- The number of groups of data: one, two, or more than two
- The relationship stated in the alternative hypothesis, H_1: not equal to or inequality (less than, greater than)
- The type of variable (population parameter): numerical (mean) or categorical (proportion)

Number of Groups

One group of hypothesis tests, known as one-sample tests, are of limited practical use because if you are interested in examining the value of a population parameter, you can usually use one of the confidence interval estimate methods that Chapter 6 discusses. The more useful two-sample tests, examining the differences between two groups, form part of the solution to the WORKED-OUT PROBLEMS of Sections 8.1 through 8.3. (Chapter 9 discusses tests for more than two groups.)

Relationship Stated in the Alternative Hypothesis, H_1

Alternative hypotheses can be stated either using the not-equal sign, as in H_1: $\mu_1 \neq \mu_2$; or by using an inequality, such as H_1: $\mu_1 > \mu_2$. You use a **two-tail test** for alternative hypotheses that use the not-equal sign and use a **one-tail test** for alternative hypotheses that contain an inequality.

One-tail and two-tail test procedures are very similar and differ mainly in the way they use critical values to determine the region of rejection. Throughout this book, two-tail hypothesis tests are featured. One-tail tests are not further discussed in this book, although the WORKED-OUT PROBLEM 8 in Chapter 8 illustrates one possible use for such tests.

Type of Variable

The type of variable, numerical or categorical, also influences the choice of hypothesis test used. For a numerical variable, the test might examine the population mean or the differences among the means, if two or more groups are used. For a categorical variable, the test might examine the population proportion or the differences among the population proportions if two or more groups are used. Chapter 8 discusses tests for two groups for each type of variable. Chapter 9 discusses tests involving more than two groups for each type of variable.

One-Minute Summary

Hypotheses:

- Null hypothesis
- Alternative hypothesis

Types of errors:

- Type I error
- Type II error

Hypothesis testing approach:

- Test statistic
- p-value

Hypothesis test relationship:

- One-tail test
- Two-tail test

Test Yourself

1. A type II error is committed when:
   a.   you reject a null hypothesis that is true
   b.   you don't reject a null hypothesis that is true
   c.   you reject a null hypothesis that is false
   d.   you don't reject a null hypothesis that is false

2. A type I error is committed when:
   a.   you reject a null hypothesis that is true
   b.   you don't reject a null hypothesis that is true
   c.   you reject a null hypothesis that is false
   d.   you don't reject a null hypothesis that is false

3. Which of the following is an appropriate null hypothesis?
   a.   The difference between the means of two populations is equal to 0.
   b.   The difference between the means of two populations is not equal to 0.
   c.   The difference between the means of two populations is less than 0.
   d.   The difference between the means of two populations is greater than 0.

4. Which of the following is not an appropriate alternative hypothesis?
   a.   The difference between the means of two populations is equal to 0.
   b.   The difference between the means of two populations is not equal to 0.
   c.   The difference between the means of two populations is less than 0.
   d.   The difference between the means of two populations is greater than 0.

5. The power of a test is the probability of:
   a.   rejecting a null hypothesis that is true
   b.   not rejecting a null hypothesis that is true
   c.   rejecting a null hypothesis that is false
   d.   not rejecting a null hypothesis that is false

6. If the p-value is less than α in a two-tail test:
    a.   the null hypothesis should not be rejected
    b.   the null hypothesis should be rejected
    c.   a one-tail test should be used
    d.   no conclusion can be reached

7. A test of hypothesis has a type I error probability (α) of 0.01. Which of the following is correct?
    a.   If the null hypothesis is true, you don't reject it 1% of the time.
    b.   If the null hypothesis is true, you reject it 1% of the time.
    c.   If the null hypothesis is false, you don't reject it 1% of the time.
    d.   If the null hypothesis is false, you reject it 1% of the time.

8. Which of the following statements is not true about the level of significance in a hypothesis test?
    a.   The larger the level of significance, the more likely you are to reject the null hypothesis.
    b.   The level of significance is the maximum risk you are willing to accept in making a type I error.
    c.   The significance level is also called the α level.
    d.   The significance level is another name for a type II error.

9. If you reject the null hypothesis when it is false, then you have committed:
    a.   a type II error
    b.   a type I error
    c.   no error
    d.   a type I error and a type II error

10. The probability of a type _____ error is also called "the level of significance."

11. The probability of a type I error is represented by the symbol _____.

12. The value that separates a rejection region from a nonrejection region is called the _____.

13. Which of the following is an appropriate null hypothesis?
    a.   The mean of a population is equal to 100.
    b.   The mean of a sample is equal to 50.
    c.   The mean of a population is greater than 100.
    d.   All of the above.

14. Which of the following is an appropriate alternative hypothesis?
    a.   The mean of a population is equal to 100.
    b.   The mean of a sample is equal to 50.
    c.   The mean of a population is greater than 100.
    d.   All of the above.

Answer True or False:

15. For a given level of significance, if the sample size is increased, the power of the test will increase.

16. For a given level of significance, if the sample size is increased, the probability of committing a type I error will increase.

17. The statement of the null hypothesis always contains an equality.

18. The larger the p-value, the more likely you are to reject the null hypothesis.

19. The statement of the alternative hypothesis always contains an equality.

20. The smaller the p-value, the more likely you are to reject the null hypothesis.

Answers to Test Yourself

1. d
2. a
3. a
4. a
5. c
6. b
7. b
8. d
9. c
10. I

11. α
12. critical value
13. a
14. c
15. True
16. False
17. True
18. False
19. False
20. True

References

1. Berenson, M. L., D. M. Levine, K. A. Szabat, and D. Stephan. *Basic Business Statistics: Concepts and Applications*, 15th edition. Hoboken, NJ: Pearson Education, 2023.

2. Levine, D. M., D. Stephan, and K. A. Szabat. *Statistics for Managers Using Microsoft Excel*, 9th edition. Hoboken, NJ: Pearson Education, 2021.

Hypothesis Testing: Z and t Tests

Chapter 8 introduces the fundamentals of hypothesis testing. This chapter discusses hypothesis tests that involve two groups, also known as two-sample tests. You will learn to use:

- The hypothesis test that examines the differences between the proportions of two groups

- The hypothesis test that examines the differences between the means of two groups

You will also learn how to evaluate the statistical assumptions about your variables that need to be true in order to use these tests and what to do if the assumptions do not hold.

8.1   Test for the Difference Between Two Proportions

CONCEPT   Hypothesis test that analyzes differences between two groups by examining the differences in sample proportions of the two groups.

INTERPRETATION   The sample proportion for each group is the number of successes in the group sample divided by the group's sample size. Sample proportions for both groups are needed as the test statistic is based on the difference in the sample proportion of the two groups. With a sufficient sample size in each group, the sampling distribution of the difference between the two proportions approximately follows a normal distribution.

WORKED-OUT PROBLEM 1   Businesses use a method called *A/B testing* to test different web page designs to see if one design is more effective than another. For one company, designers were interested in the effect of modifying the call-to-action button on the home page. Every visitor to the company's home page was randomly shown either the original call-to-action button (the control) or the new variation.

Designers measured success by the download rate: the number of people who downloaded the file divided by the number of people who saw that particular call-to-action button. Results of the experiment yielded the following data.

		Call-to-Action Button		
		Original	New	Total
Download	Yes	351	451	802
	No	3,291	3,105	6,396
	Total	3,642	3,556	7,198

Of the 3,642 people who used the original call-to-action button, 351 downloaded the file, for a proportion of 0.0964. Of the 3,556 people who used the new call-to-action button, 451 downloaded the file, for a proportion of 0.1268.

Because the number of downloads triggered by the original and new call-to-action buttons is large (351 and 451) and the number of did-not-downloads from the original and new call-to-action buttons is also large (3,291 and 3,105), the sampling distribution for the difference between the two proportions is approximately normally distributed. The null and alternative hypotheses are as follows:

H_0: $\pi_1 = \pi_2$ (No difference in the proportion of downloads triggered by the original and new call-to-action buttons)

H_1: $\pi_1 \neq \pi_2$ (There is a difference in the proportion of downloads triggered by the original and new call-to-action buttons)

Using the critical value approach with a level of significance of 0.05, the lower tail area is 0.025, and the upper tail area is 0.025. Using the cumulative normal distribution table (Table C.1), the lower critical value of 0.025 corresponds to a *Z* value of −1.96, and an upper critical value of 0.025 (cumulative area of 0.975) corresponds to a *Z* value of +1.96, as Figure 8.1 shows.

FIGURE 8.1

Rejection and
nonrejection
regions
when level of
significance
is 0.05

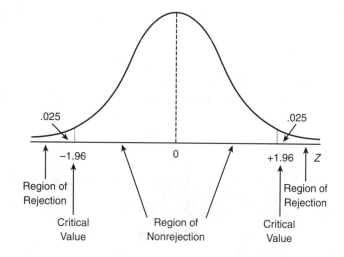

Given these rejection regions, you will reject H_0 if $Z < -1.96$ or if $Z > +1.96$; otherwise, you will not reject H_0. The Figure 8.2 spreadsheet results for WORKED-OUT PROBLEM 1 show that the Z test statistic is -4.6584. Because $Z = -4.6584$ is less than the lower critical value of -1.96, you reject the null hypothesis. You conclude that evidence exists of a difference in the proportion of downloads triggered by the original and new call-to-action buttons.

FIGURE 8.2

Spreadsheet
Z test for the
proportion
results for
WORKED-OUT
PROBLEM 1

	A	B
1	Z Test for Differences in Two Proportions	
2		
3	Data	
4	Hypothesized Difference	0
5	Level of Significance	0.05
6	Group 1	
7	Number of Successes	351
8	Sample Size	3291
9	Group 2	
10	Number of Successes	451
11	Sample Size	3105
12		
13	Intermediate Calculations	
14	Group 1 Proportion	0.1067
15	Group 2 Proportion	0.1452
16	Difference in Two Proportions	-0.0386
17	Average Proportion	0.1254
18	Z Test Statistic	-4.6584
19		
20	Two-Tail Test	
21	Lower Critical Value	-1.9600
22	Upper Critical Value	1.9600
23	p-Value	0.0000
24	Reject the null hypothesis	

WORKED-OUT PROBLEM 2   You decide to use the *p*-value approach to hypothesis testing for the previous problem. Figure 8.2 spreadsheet results show that the *p*-value is 0.0000. This means that the probability of obtaining a *Z* value less than −4.6584 or greater than +4.6584 is virtually zero (0.0000). Because the *p*-value is less than the level of significance α = 0.05, you reject the null hypothesis. You conclude that evidence exists of a difference in the proportion of downloads triggered by the original and new call-to-action buttons. The new call-to-action button is more likely to result in downloads of the file.

spreadsheet solution

Z Test for the Difference in Two Proportions

Chapter 8 Z Two Proportions contains the *Z* test for the difference in two proportions that WORKED-OUT PROBLEMS 1 and 2 uses to calculate the results of the *Z* test of two proportions.

Best Practices

Use the **NORM.S.INV(P<X)** function to calculate a critical value of the normal distribution, where *P<X* is the area under the curve that is less than *X*.

Use the **NORM.S.DIST(*Z value*, TRUE)** function to calculate the cumulative normal probability of less than the *Z* test statistic.

How-Tos

For other problems, change the hypothesized difference and level of significance in cells B4 and B5, as well as the number of successes and sample size for each group in cells B7, B8, B10, and B11.

For the two-tail test, compute the *p*-value by multiplying 2 by the expression (1 minus the absolute value of the NORM.S.DIST value).

WORKED-OUT PROBLEM 3   A famous health care experiment investigated the effectiveness of aspirin in the reduction of the incidence of heart attacks. In this experiment, 22,071 male U.S. physicians were randomly assigned to either a group that was given one 325 mg buffered aspirin tablet every other day or a group that was given a placebo (which contained no active ingredients).

The following table summarizes the results of the experiment. Of 11,037 physicians taking aspirin, 104 suffered heart attacks during the five-year period of the study. Of 11,034 physicians who were assigned to a group that took a placebo every other day, 189 suffered heart attacks during the five-year period of the study.

		Study Group		
		Aspirin	Placebo	Totals
Results	Heart attack	104	189	293
	No heart attack	10,933	10,845	21,778
	Totals	11,037	11,034	22,071

For this experiment, you establish the null and alternative hypotheses as:

H_0: $\pi_1 = \pi_2$ (No difference exists in the proportion of heart attacks between the group that was given aspirin and the group that was given the placebo.)

H_1: $\pi_1 \neq \pi_2$ (A difference exists in the proportion of heart attacks between the two groups.)

Given the null hypothesis, the number of heart attacks is the number of successes for this problem. This illustrates that the number of successes can represent a negative, or "unsuccessful," real-world outcome.

Using the **Chapter 8 Z Two Proportions** spreadsheet with the health care experiment data and using a level of significance of $\alpha = 0.05$ generates the following results area:

17	Average Proportion	0.0133
18	Z Test Statistic	-5.0014
19		
20	Two-Tail Test	
21	Lower Critical Value	-1.9600
22	Upper Critical Value	1.9600
23	p-Value	0.0000
24	Reject the null hypothesis	

Using the critical value approach, because $Z = -5.00$ is less than the lower critical value of -1.96 at the 0.05 level of significance, you reject the null hypothesis. (Likewise, using the *p*-value approach, because the *p*-value is less than the level of significance [0.05], you reject the null hypothesis.) You conclude that evidence exists of a difference in the proportion of doctors who have had heart attacks between those who took the aspirin and those who did not take the aspirin. The group who took the aspirin had a significantly lower proportion of heart attacks during the five-year experiment.

equation blackboard (optional)

interested in math?

WORKED-OUT PROBLEMS 1 through 3 use the *Z* test for the difference between two proportions. You need the subscripted symbols for the number of successes, *X*, the sample sizes, n_1 and n_2, sample proportions, p_1 and p_2, and population proportions, π_1 and π_2, as well as the symbol for the pooled estimate of the population proportion, $\bar{p}$, to calculate the *Z* test statistic.

To write the *Z* test statistic equation, you first define the symbols for the pooled estimate of the population proportion and the sample proportions for the two groups:

$$\bar{p} = \frac{X_1 + X_2}{n_1 + n_2} \quad p_1 = \frac{X_1}{n_1} \quad p_2 = \frac{X_2}{n_2}$$

Next, you use $\bar{p}$, p_1, and p_2 along with the symbols for the sample sizes and population proportion to form the equation for the *Z* test for the difference between two proportions:

$$Z = \frac{(p_1 - p_2) - (\pi_1 - \pi_2)}{\sqrt{\bar{p}(1 - \bar{p})\left(\dfrac{1}{n_1} + \dfrac{1}{n_2}\right)}}$$

As an example, the calculations for WORKED-OUT PROBLEM 1 are:

$$p_1 = \frac{X_1}{n_1} = \frac{351}{3,642} = 0.0964 \quad p_2 = \frac{X_2}{n_2} = \frac{451}{3,556} = 0.1268$$

and

$$\bar{p} = \frac{X_1 + X_2}{n_1 + n_2} = \frac{351 + 451}{3,642 + 3,556} = \frac{802}{7,198} = 0.1114$$

So that:

$$Z_{STAT} = \frac{(0.0964 - 0.1268) - (0)}{\sqrt{0.1114(1 - 0.1114)\left(\dfrac{1}{3,642} + \dfrac{1}{3,556}\right)}}$$

$$= \frac{-0.0304}{\sqrt{(0.09899)(0.0005557)}} = \frac{-0.0304}{\sqrt{0.000055}}$$

$$= \frac{-0.0304}{0.00742}$$

$$= -4.10$$

With $\alpha = 0.05$, you reject H_0 if $Z < -1.96$ or if $Z > +1.96$; otherwise, do not reject H_0. Because $Z = -4.10$ is less than the lower critical value of -1.96, you reject the null hypothesis.

8.2 Test for the Difference Between the Means of Two Independent Groups

CONCEPT   Hypothesis test that analyzes differences between two groups by determining whether a significant difference exists in the population means of two populations or groups.

INTERPRETATION   Statisticians distinguish between using two independent groups and using two related groups when performing this type of hypothesis test. With related groups, either the observations are matched according to a relevant characteristic or repeated measurements of the same items are taken. For studies involving two independent groups, the most common test of hypothesis used is the pooled-variance t test.

Pooled-Variance *t* Test

CONCEPT   The hypothesis test for the difference between the population means of two independent groups that combines, or "pools," the sample variance of each group into one estimate of the variance common in the two groups.

INTERPRETATION   For this test, the test statistic is based on the difference in the sample means of the two groups, and the sampling distribution for the difference in the two sample means approximately follows the t distribution.

In a pooled-variance *t* test, the null and alternate hypotheses are:

$H_0: \mu_1 = \mu_2$ (The two population means are equal.)

and

$H_1: \mu_1 \neq \mu_2$ (The two population means are not equal.)

Super Bowl Ads

WORKED-OUT PROBLEM 4   You wish to further analyze the data in **Super Bowl Ads** that is first used in Problem 4 in Chapter 2. This file contains the average ratings of 57 ads from the 2021 NFL Super Bowl broadcast. For this analysis, you want to determine if there is a difference in the first half and second half mean ad ratings, using the pooled-variance *t* test and a level of significance of 0.05.

Figure 8.3 spreadsheet results for this pooled-variance *t* test show that the *t* statistic is 1.3627, and the *p*-value is 0.1785. Because *t* = 1.3627 < 2.0040 or because the *p*-value, 0.1785, is greater than α = 0.05, you do not reject the null hypothesis. You conclude that you have insufficient evidence of a significant difference between the first half and second half mean ad ratings.

FIGURE 8.3

Spreadsheet for the pooled-variance *t* test for WORKED-OUT PROBLEM 4

	A	B
1	Pooled-Variance t Test for Differences in Two Means	
2	(assumes equal population variances)	
3	Data	
4	Hypothesized Difference	0
5	Level of Significance	0.05
6	Population 1 Sample	
7	Sample Size	28
8	Sample Mean	5.7893
9	Sample Standard Deviation	0.7020
10	Population 2 Sample	
11	Sample Size	29
12	Sample Mean	5.5345
13	Sample Standard Deviation	0.7093
14		
15	Intermediate Calculations	
16	Population 1 Sample Degrees of Freedom	27
17	Population 2 Sample Degrees of Freedom	28
18	Total Degrees of Freedom	55
19	Pooled Variance	0.4980
20	Standard Error	0.1870
21	Difference in Sample Means	0.2548
22	t Test Statistic	1.3627
23		
24	Two-Tail Test	
25	Lower Critical Value	-2.0040
26	Upper Critical Value	2.0040
27	p-Value	0.1785
28	Do not reject the null hypothesis	

WORKED-OUT PROBLEM 5   The Adglow.com blog post "E-Commerce: Men spend more than women" reports that women buy a mean of 7.1 Internet purchases per year, while men buy a mean of 5.4 Internet purchases. Suppose these findings were based on a study that consisted of 25 men and 30 women and that the standard deviation for the women's purchases was 3.2 and the standard deviation for the men's purchases was 2.3. Using the pooled-variance *t* test and a level of significance of 0.05, is there evidence of a difference in the mean purchases per year between men and women?

Spreadsheet results for this pooled-variance *t* test are as follows.

	A	B
1	**Pooled-Variance *t* Test for Differences in Two Means**	
2	(assumes equal population variances)	
3	**Data**	
4	**Hypothesized Difference**	0
5	**Level of Significance**	0.05
6	**Population 1 Sample**	
7	**Sample Size**	30
8	**Sample Mean**	7.1
9	**Sample Standard Deviation**	3.2
10	**Population 2 Sample**	
11	**Sample Size**	25
12	**Sample Mean**	5.4
13	**Sample Standard Deviation**	2.3
14		
15	Intermediate Calculations	
16	Population 1 Sample Degrees of Freedom	29
17	Population 2 Sample Degrees of Freedom	24
18	Total Degrees of Freedom	53
19	Pooled Variance	7.9985
20	Standard Error	0.7659
21	Difference in Sample Means	1.7
22	*t* Test Statistic	2.2197
23		
24	**Two-Tail Test**	
25	**Lower Critical Value**	-2.0057
26	**Upper Critical Value**	2.0057
27	*p-Value*	0.0307
28	**Reject the null hypothesis**	

The results show that the *t* statistic is 2.2197, and the *p*-value is 0.0307. Because $t = 2.2197 > 2.0057$, or the *p*-value is less than $\alpha = 0.05$, you reject the null hypothesis. You conclude that you have evidence of a difference in the mean number of Internet purchases per year for women and men.

spreadsheet solution

Pooled-Variance *t* Test for the Differences in Two Means

Chapter 8 Pooled-Variance T with Unsummarized Data contains the spreadsheet that WORKED-OUT PROBLEM 4 uses to perform a pooled-variance *t* test for the difference in two means with the unsummarized data found in columns E and F of the spreadsheet (columns not shown in Figure 8.3).

Chapter 8 Pooled-Variance T with Sample Statistics contains the spreadsheet that WORKED-OUT PROBLEM 5 uses to perform a pooled-variance *t* test for the difference in two means with sample statistics. The spreadsheet is identical to the other spreadsheet except for the contents of cells B7 through B9 and B11 and B13, which contain numbers and not simple formulas that calculate those numbers using the unsummarized data, and for the empty columns E and F.

Best Practices

Use the **T.INV.2T**(*level of confidence, degrees of freedom*) function to calculate the upper critical value of the *t* distribution. Precede the same function with a minus sign to calculate the lower critical value of the *t* distribution.

Use the **T.DIST.2T**(*absolute value of the t test statistic, total degrees of freedom*) function to calculate a probability associated with the *t* distribution.

How-Tos

For other problems that use unsummarized data, enter new unsummarized data in columns E and F, replacing the data that are already there. For other problems that use sample statistics, enter the sample size, sample mean, and sample standard deviations in cells B7 through B9 (population 1 sample) and B11 and B13 (population 2 sample). For other problems of either type, change the hypothesized difference in cell B4 and the level of significance in cell B5, as necessary.

Analysis ToolPak Tip ATT3 in Appendix E explains how to use Analysis ToolPak as a second way to perform a pooled-variance *t* test for the difference in two means using unsummarized data.

equation
blackboard
(optional)

interested
in
math?

WORKED-OUT PROBLEMS 4 and 5 use the pooled-variance t test for the difference between the population means of two independent groups. You need the subscripted symbols for the sample means, $\bar{X}_1$ and $\bar{X}_2$, the sample sizes for each of the two groups, n_1 and n_2, and the population means, μ_1 and μ_2, along with the symbol for the pooled estimate of the variance, S_p^2, to calculate the t test statistic.

To write the equation for the t test statistic, you first define the symbols for the equation for the pooled estimate of the population variance:

$$S_p^2 = \frac{(n_1 - 1)S_1^2 + (n_2 - 1)S_2^2}{(n_1 - 1) + (n_2 - 1)}$$

You next use S_p^2, which you just defined, and the symbols for the sample means, the population means, and the sample sizes to form the equation for the pooled-variance t test for the difference between two means:

$$t = \frac{\left(\bar{X}_1 - \bar{X}_2\right) - \left(\mu_1 - \mu_2\right)}{\sqrt{S_p^2 \left(\frac{1}{n_1} + \frac{1}{n_2}\right)}}$$

The calculated test statistic t follows a t distribution with $n_1 + n_2 - 2$ degrees of freedom.

As an example, the calculations for WORKED-OUT PROBLEM 4 are:

$$S_p^2 = \frac{(n_1 - 1)S_1^2 + (n_2 - 1)S_2^2}{(n_1 - 1) + (n_2 - 1)}$$

$$= \frac{27(0.4928) + 28(0.5031)}{27 + 28} = 0.4980$$

Using 0.4980 as the value for S_p^2 in the original equation:

$$t = \frac{\left(\bar{X}_1 - \bar{X}_2\right) - \left(\mu_1 - \mu_2\right)}{\sqrt{S_p^2 \left(\frac{1}{n_1} + \frac{1}{n_2}\right)}}$$

produces:

$$t = \frac{(5.7893 - 5.5345) - 0}{\sqrt{0.4980\left(\frac{1}{28} + \frac{1}{29}\right)}}$$

$$t = \frac{0.2548}{\sqrt{0.4980(0.0702)}} = \frac{0.2548}{\sqrt{0.03496}}$$

$$t = +1.3627$$

Using the $\alpha = 0.05$ level of significance, with $28 + 29 - 2 = 55$ degrees of freedom, the critical value of t is 2.0040 (0.025 in the upper tail of the t distribution). Because $t = +1.3627 < 2.0040$, you do not reject H_0.

Pooled-Variance *t* Test Assumptions

In testing for the difference between the means, you assume that the two populations from which the two independent samples have been selected are normally distributed with equal variances. When that assumption holds, the pooled-variance t test is valid even if there is a moderate departure from normality, as long as the sample sizes are large.

You can check the assumption of normality by comparing boxplots for the two groups. Boxplots for the two groups of Super Bowl ad ratings appear as part of WORKED-OUT PROBLEM 15 in Chapter 3 and show that the first half ad ratings are slightly right-skewed. However, given the sample sizes in each of the two groups, you conclude that any departure from the normality assumption will not seriously affect the validity of the t test.

If the data in each group cannot be assumed to be from normally distributed populations, you use a nonparametric procedure, such as the Wilcoxon rank sum test, that does not depend on that assumption. The pooled-variance t test also assumes that the population variances are equal. If this assumption cannot be made, you cannot use the pooled-variance t test, and you use the separate-variance t test (see references 1 and 2).

8.3 The Paired *t* Test

CONCEPT   A hypothesis test for the difference between two groups for situations in which the data for the two groups are *related and not independent*. In this test, the variable of interest is the difference between related pairs of values in the two groups rather than the paired values themselves.

INTERPRETATION   There are two situations in which the values from two groups will be related and not independent.

In the first case, a researcher has paired, or matched, the values under study according to some other variable. For example, in testing whether a new drug treatment lowers blood pressure, a sample of patients could be paired according to their blood pressure at the beginning of the study. In this way, if there were two patients in the study who each had a diastolic blood pressure of 140, one would be randomly assigned to the group that will take the new drug and the other to the group that will not take the new drug.

Assigning patients in this manner means that the researcher will not have to be concerned about differences in the initial blood pressures of the patients that form the two groups. This, in turn, means that test results will better reflect the effect of the new drug being tested.

In the second case, a researcher obtains two sets of measurements from the same items or individuals. This approach is based on the premise that the same items or individuals will behave alike if treated alike. This, in turn, allows the researcher to assert that any differences between two sets of measurements are due to what is under study. For example, when performing an experiment on the effect of a diet drug, the researcher could take one measurement from each participant just prior to starting the drug and one just after the end of a specified time period.

In both cases, the variable of interest can be stated algebraically as follows:

Difference (D) = Related value in sample 1 – Related value in sample 2

With related groups and a numerical variable of interest, the null hypothesis is that no difference exists in the population means of the two related groups, and the alternative hypothesis is that there is a difference in the population means of the two related groups. Using the symbol μ_D to represent the difference between the population means, the null and alternative hypotheses can be expressed as follows:

$H_0: \mu_D = 0$

and:

$H_1: \mu_D \neq 0$

To decide whether to reject the null hypothesis, you use a paired t test.

WORKED-OUT PROBLEM 6   At the 0.05 level of significance, you want to determine whether there is a difference between the cost of an inexpensive restaurant meal and a McDonald's McMeal in 25 world cities. You collect the following data and calculate differences:

World Meal Costs

City	Inexpensive Restaurant Meal	McDonald's McMeal
Bangalore, India	2.70	4.05
Bangkok, Thailand	2.07	5.92
Beijing, China	4.64	5.41
Berlin, Germany	11.60	9.28
Budapest, Hungary	7.95	5.63
Buenos Aires, Argentina	7.68	6.94
Dubai, United Arab Emirates	9.53	7.62
Hong Kong, Hong Kong	7.71	5.14
Istanbul, Turkey	4.51	3.95
Johannesburg, South Africa	10.06	4.02
Kampala, Uganda	1.70	7.78
Lima, Peru	3.15	4.84
London, United Kingdom	20.32	8.13
Melbourne, Australia	14.54	8.73
Mexico City, Mexico	7.33	5.87
Moscow, Russia	10.99	4.81
Nairobi, Kenya	4.52	6.33
New York, NY, United States	20.00	9.00
Paris, France	17.39	10.44
Rome, Italy	17.39	9.28
Sao Paulo, Brazil	5.59	5.96
Stockholm, Sweden	13.71	9.14
Toronto, Canada	15.85	9.23
Warsaw, Poland	8.86	5.57
Zurich, Switzerland	26.87	16.12

Source: Data extracted from "Prices from City," https://www.numbeo.com.

Because the two sets of meal costs are from the same world cities, the two sets of measurements are related, and only the differences between the cost of inexpensive restaurant meals and McDonald's McMeals are tested using a paired *t* test.

The Figure 8.4 spreadsheet results on page 171 show that the *t* statistic is 3.3045, and the *p*-value is 0.0030. Because the *p*-value, 0.0030, is less than $\alpha = 0.05$ (or because $t = 3.3045 > 2.0639$), you reject the null hypothesis. You conclude that a difference exists between the cost of an inexpensive restaurant meal and a McDonald's McMeal in 25 world cities.

FIGURE 8.4

Spreadsheet
or paired
t test for
WORKED-OUT
PROBLEM 6

	A	B
1	**Paired *t* Test of Inexpensive Meals and McMeals**	
2		
3	**Data**	
4	**Hypothesized Mean Difference**	0
5	**Level of significance**	0.05
6		
7	**Intermediate Calculations**	
8	Sample Size	25
9	DBar	3.0988
10	Degrees of Freedom	24
11	S_D	4.6887
12	Standard Error	0.9377
13	*t* Test Statistic	3.3045
14		
15	**Two-Tail Test**	
16	Lower Critical Value	-2.0639
17	Upper Critical Value	2.0639
18	*p*-Value	0.0030
19	Reject the null hypothesis	

spreadsheet solution

Paired *t* Test

Chapter 8 Paired T contains the spreadsheet that WORKED-OUT PROBLEM 6 uses to perform the paired *t* test for the difference in two means for unsummarized data.

Best Practices

Use the **DEVSQ**(*column of differences*) function to compute the sum of the squares of the differences between each set of paired values and the mean difference to help calculate the sample standard deviation.

Use the **T.INV.2T**(*level of confidence, degrees of freedom*) function to calculate the upper critical value of the *t* distribution. Precede the same function with a minus sign to calculate the lower critical value of the *t* distribution.

Use the **T.DIST.2T**(*absolute value of the t test statistic, total degrees of freedom*) function to calculate a probability associated with the *t* distribution.

How-Tos

Change the hypothesized mean difference and level of significance in cells B4 and B5 to experiment with this spreadsheet.

Use the similar **Chapter 8 Paired T Advanced** spreadsheet to enter new unsummarized data of any sample size up to 50 or less to be entered in its columns E through G.

Analysis ToolPak Tip ATT4 in Appendix E describes the use of Analysis ToolPak as a second way to perform a paired *t* test for the difference in two means.

Advanced Technique ADV5 in Appendix E explains how the **Chapter 8 Paired T Advanced** spreadsheet uses the ISBLANK function to display difference values only for those rows that contain data. This section also explains how to modify the spreadsheet for data sets that have more than 50 rows of values.

equation blackboard (optional)

interested in math?

WORKED-OUT PROBLEM 6 uses the equation for the paired *t* test. You need the symbols for the sample size, n, the difference between the population means, μ_D, the sample standard deviation, S_D, and the subscripted symbol for the differences in the paired values, D_i, all previously introduced, and the symbol for the mean difference, $\bar{D}$, to calculate the *t* test statistic.

To write the *t* test statistic equation, you first define the symbols for the equation for the mean difference, $\bar{D}$:

$$\bar{D} = \frac{\sum_{i=1}^{n} D_i}{n}$$

You next use $\bar{D}$, the symbols for the sample size, and the differences in the paired values to form the equation for the sample standard deviation, S_D:

$$S_D = \sqrt{\frac{\sum_{i=1}^{n} (D_i - \bar{D})^2}{n-1}}$$

Finally, you assemble $\bar{D}$ and S_D and the remaining symbols to form the equation for the paired t test for the difference between two means:

$$t = \frac{\bar{D} - \mu_D}{\frac{S_D}{\sqrt{n}}}$$

The test statistic t follows a t distribution with $n - 1$ degrees of freedom.

As an example, the calculations for determining the t test statistic for WORKED-OUT PROBLEM 6 are:

$$\bar{D} = \frac{\sum_{i=1}^{n} D_i}{n}$$

$$\bar{D} = \frac{77.47}{25} = 3.0988$$

Using that value, $S_D = 4.6887$ (calculation not shown). Substituting values results in the t value -1.0560:

$$t = \frac{\bar{D} - \mu_D}{\frac{S_D}{\sqrt{n}}} = \frac{3.0988 - 0}{\frac{4.6887}{\sqrt{25}}} = 3.3045$$

Using the level of significance $\alpha = 0.05$, with $25 - 1 = 24$ degrees of freedom, the critical value of t is 2.0639 (0.025 in the upper tail of the t distribution). Because $t = -3.3045 > 2.0639$, you reject the null hypothesis, H_0.

WORKED-OUT PROBLEM 7   You seek to determine, using a level of significance of $\alpha = 0.05$, whether differences exist in monthly sales between the new package design and the old package design of a laundry stain remover. The new package was test marketed over a period of one month in a sample of supermarkets in a particular city.

A random sample of 10 pairs of supermarkets was matched according to weekly sales volume and a set of demographic characteristics. The following table shows the data collected.

Supermarket

Pair	New Package	Old Package	Difference
1	458	437	21
2	519	488	31
3	394	409	−15
4	632	587	45
5	768	753	15
6	348	400	−52
7	572	508	64
8	704	695	9
9	527	496	31
10	584	513	71

Because the 10 pairs of supermarkets were matched, you use the paired *t* test. The Figure 8.5 spreadsheet results for this study show that the *t* statistic is 1.9116, and the *p*-value is 0.0882. This means that the chance of obtaining a *t* value greater than 1.91 or less than −1.91 is 0.0882, or 8.82%. Because the *p*-value, 0.0882, is greater than $\alpha = 0.05$ (or because $t = 1.9116 < 2.2622$), you do not reject the null hypothesis. You can conclude that insufficient evidence exists of a difference between the new and old package designs.

FIGURE 8.5

Spreadsheet for paired *t* test for WORKED-OUT PROBLEM 7

	A	B
1	**Paired *t* Test**	
2		
3	**Data**	
4	**Hypothesized Mean Difference**	0
5	**Level of Significance**	0.05
6		
7	**Intermediate Calculations**	
8	Sample Size	10
9	Mean Difference	22.0000
10	Degrees of Freedom	9
11	Sample Standard Deviation	36.3929
12	Standard Error	11.5085
13	*t* Test Statistic	1.9116
14		
15	**Two-Tailed Test**	
16	**Lower Critical Value**	-2.2622
17	**Upper Critical Value**	2.2622
18	***p*-Value**	0.0882
19	**Do not reject the null hypothesis**	

WORKED-OUT PROBLEM 8 Before the test marketing experiment, you might have wanted to determine whether the new package design produced *more* sales than the old package design and not just a difference in sales. Such a situation would be a good application of a one-tail test in which the alternative hypothesis is $H_1: \mu_D > 0$.

In such a case, you would use the one-tail p-value (or one-tail critical value). For the test market sales data, the one-tail p-value is 0.0441 (and the one-tail critical value is 1.8331). Because the p-value is less than $\alpha = 0.05$ (or because $t = 1.9116$ is greater than 1.8331), you reject the null hypothesis. You conclude that the mean sales from the new package design were higher than the mean sales for the old package design; this is a different conclusion than you made using the two-tail test.

Important Equations

Z Test for the Difference Between Two Proportions:

$$Z = \frac{(p_1 - p_2) - (\pi_1 - \pi_2)}{\sqrt{\bar{p}(1-\bar{p})\left(\dfrac{1}{n_1} + \dfrac{1}{n_2}\right)}}$$

Pooled-Variance t Test for the Difference Between the Population Means of Two Independent Groups:

$$t = \frac{(\bar{X}_1 - \bar{X}_2) - (\mu_1 - \mu_2)}{\sqrt{S_p^2\left(\dfrac{1}{n_1} + \dfrac{1}{n_2}\right)}}$$

Paired t Test for the Difference Between Two Means:

$$t = \frac{\bar{D} - \mu_D}{\dfrac{S_D}{\sqrt{n}}}$$

One-Minute Summary

For tests for the differences between two groups, first determine whether your data are categorical or numerical:

- If your data are categorical, use the *Z* test for the difference between two proportions.
- If your data are numerical, determine whether you have independent or related groups:
  - If you have independent groups, use the pooled-variance *t* test for the difference between two means.
  - If you have related groups, use the paired *t* test.

Test Yourself
Short Answers

1. The *t* test for the difference between the means of two independent populations assumes that the two:

    a. sample sizes are equal
    b. sample medians are equal
    c. populations are approximately normally distributed
    d. All of the above are correct.

2. In testing for differences between the means of two related populations, the null hypothesis is:

    a. $H_0: \mu_D = 2$
    b. $H_0: \mu_D = 0$
    c. $H_0: \mu_D < 0$
    d. $H_0: \mu_D > 0$

3. A researcher is curious about the effect of sleep on students' test performances. He chooses 100 students and gives each student two exams. One is given after four hours' sleep and the other after eight hours' sleep. The statistical test the researcher should use is the:

    a. *Z* test for the difference between two proportions
    b. pooled-variance *t* test
    c. paired *t* test

4. A statistics professor wanted to test whether the grades on a statistics test were the same for her morning class and her afternoon class. For this situation, the professor should use the:

    a. *Z* test for the difference between two proportions
    b. pooled-variance *t* test
    c. paired *t* test

Answer True or False:

5. The sample size in each independent sample must be the same in order to test for differences between the means of two independent populations.

6. In testing a hypothesis about the difference between two proportions, the *p*-value is computed to be 0.043. The null hypothesis should be rejected if the chosen level of significance is 0.05.

7. In testing a hypothesis about the difference between two proportions, the *p*-value is computed to be 0.034. The null hypothesis should be rejected if the chosen level of significance is 0.01.

8. In testing a hypothesis about the difference between two proportions, the Z test statistic is computed to be 2.04. The null hypothesis should be rejected if the chosen level of significance is 0.01 and a two-tail test is used.

9. The sample size in each independent sample must be the same in order to test for differences between the proportions of two independent populations.

10. When you are sampling the same individuals and taking a measurement before treatment and after treatment, you should use the paired *t* test.

11. Repeated measurements from the same individuals are an example of data collected from two related populations.

12. The pooled-variance *t* test assumes that the population variances in the two independent groups are equal.

13. In testing a null hypothesis about the difference between two proportions, the Z test statistic is computed to be 2.04. The *p*-value is 0.0207.

14. You can use a pie chart to evaluate whether the assumption of normally distributed populations in the pooled-variance *t* test has been violated.

15. If the assumption of normally distributed populations in the pooled-variance *t* test has been violated, you should use an alternative procedure such as the nonparametric Wilcoxon rank sum test.

Answers to Test Yourself Short Answers

1. c
2. b
3. c
4. b
5. False
6. True
7. False
8. False

9. False
10. True
11. True
12. True
13. False
14. False
15. True

Problems

1. Students who attend a regional university located in a small town are known to favor the local pizza restaurant. A national chain of pizza restaurants looks to conduct a survey of students to determine pizza preferences. The chain selects a sample of 220 students attending the university. Of 93 females sampled, 74 preferred the local pizza restaurant. Of 127 males sampled, 71 preferred the local pizza restaurant.

   At the 0.05 level of significance, is there evidence of a difference in the proportion of female students who prefer the local restaurant and the proportion of male students who prefer the same?

2. Churning, the loss of customers to a competitor, is especially a problem for telecommunications companies. Market researchers at a telecommunications company collect data from a sample of 5,517 customers. Of 2,741 females selected, 883 churned in the last month. Of 2,776 males selected, 873 churned in the last month.

   At the 0.05 level of significance, is there evidence of a difference in the proportion of female customers who churned and the proportion of male customers who churned?

3. Market researchers at the company in Problem 2 reviewed the same sample, looking for a pattern between churning and paperless billing. Of the 3,725 customers who used paperless billing, 1,358 churned in the past month. Of the 1,792 customers who did not use paperless billing, 398 churned in the last month.

   a. At the 0.05 level of significance, is there evidence of a difference in the proportion of customers who churned between customers who used paperless billing and those customers who did not?

   b. Suppose that the market researchers collected another sample. Of 50 customers who used paperless billing, 18 churned in the past month. Of 50 customers who did not use paperless billing, 11 churned in the past month. At the 0.05 level of significance, is there evidence of a difference in the proportion of customers who churned between customers who used paperless billing and those customers who did not?

   c. Compare the results of parts (a) and (b). What effect did the sample size have on the results?

4. When people make estimates, they are influenced by anchors to their estimates. A study was conducted in which students were asked to estimate the number of calories in a cheeseburger. One group was asked to do this after thinking about a calorie-laden cheesecake. A second group was asked to do this after thinking about an organic fruit salad. The mean number of calories estimated in a cheeseburger was 780 for the group that thought

about the cheesecake and 1,041 for the group that thought about the organic fruit salad. (Source: "Drilling Down, Sizing Up a Cheeseburger's Caloric Heft," *The New York Times*, 4 October 2010, p. B2.)

Suppose that the study was based on a sample of 20 people who thought about the cheesecake first and 20 people who thought about the organic fruit salad first. Also suppose that the standard deviation of the number of calories in the cheeseburger was 128 for the people who thought about the cheesecake first and 140 for the people who thought about the organic fruit salad first.

    a.   State the null and alternative hypotheses if you want to determine whether there is a difference in the mean estimated number of calories in the cheeseburger for the people who thought about the cheesecake first and for the people who thought about the organic fruit salad first.

    b.   At the 0.05 level of significance, is there evidence of a difference in the mean estimated number of calories in the cheeseburger for the people who thought about the cheesecake first and those who thought about the organic fruit salad first?

5. You wonder if males and females spend different amounts of time per day accessing the Internet with a mobile device. You select a sample of 60 friends and family members, 30 males and 30 females, and collect the time each spends per day (in minutes) accessing the Internet with a mobile device. The results showed that males spent a mean of 188.63 minutes, with a standard deviation of 49.73 minutes, and females spent a mean of 159.31 minutes, with a standard deviation of 34.99 minutes.

Assume that the variances in the population of times spent per day accessing the Internet via a mobile device are equal. At the 0.05 level of significance, is there evidence of a difference between males and females in the mean time spent per day accessing the Internet via a mobile device?

6. Market researchers know that users are very concerned about smartphone battery life. Researchers conducted an experiment in which the battery life of a Samsung Galaxy phone was compared to that of an iPhone. They found that a Samsung Galaxy phone had a mean life of 11.58 hours, and an iPhone had a mean life of 11.16 hours. (Data extracted from "Smartphones with the Best Battery Life," https://bit.ly/3D8EnEy.)

Suppose that the survey was conducted on 36 smartphones of each brand, and the standard deviation was 3.5 hours for a Samsung Galaxy phone and 3.2 hours for an iPhone. At the 0.05 level of significance, is there evidence of a difference in the mean battery life between a Samsung Galaxy phone and an iPhone?

7. Have you ever wondered how prices vary between supermarkets? The data in the following table compare the prices of a variety of items ordered online from Aldi, the Germany-based discount supermarket, and Publix, a U.S. regional supermarket chain, on a day in February 2021.

Aldi Publix Prices

Product	Aldi	Publix
Bananas/lb.	0.39	0.75
Medium avocado	0.55	1.37
Organic strawberries/lb.	3.98	3.69
Organic blueberries/6 oz.	2.75	5.53
Butternut squash	1.09	1.65
Granulated sugar/4 lb.	2.19	3.49
Corn oil/48 oz.	3.69	4.41
Seedless watermelon	3.85	5.53
Cantaloupe	2.19	3.87
Dark chocolate morsels/12 oz.	2.34	4.75
Chopped walnuts/4 oz.	1.43	3.31

At the 0.05 level of significance, is there evidence of a difference between the mean price of the items ordered from Aldi and the mean prices of the items ordered from Publix?

8. Multiple myeloma, or blood plasma cancer, is characterized by increased blood vessel formulation (angiogenesis) in the bone marrow that is a prognostic factor in survival. One treatment approach used for multiple myeloma is stem cell transplantation using the patient's own stem cells. The following data represent the bone marrow microvessel density for patients who had a complete response to the stem cell transplant, as measured by blood and urine tests. The measurements were taken immediately prior to the stem cell transplant and at the time of the complete response:

Myeloma

Patient	Before	After
1	158	284
2	189	214
3	202	101
4	353	227
5	416	290
6	426	176
7	441	290

Source: Extracted from S. V. Rajkumar, R. Fonseca, T. E. Witzig, M. A. Gertz, and P. R. Greipp, "Bone Marrow Angiogenesis in Patients Achieving Complete Response After Stem Cell Transplantation for Multiple Myeloma," *Leukemia*, 1999, 13, pp. 469–472.

At the 0.05 level of significance, is there evidence of a difference in the mean bone marrow microvessel density before the stem cell transplant and after the stem cell transplant?

9. Have you wondered whether there is a difference in taste between coffee brewed in a coffeepot and coffee brewed in a K-cup device? Ten friends rate each coffee type on four characteristics—taste, aroma, richness, and acidity—using a 7-point scale for each attribute in which 1 is extremely unpleasing and 7 is extremely pleasing. You sum the four values for each rating and store the total ratings in **Coffee Brew Ratings**, which contains the following data:

Coffee Brew Ratings

Reviewer	Coffeepot	K-Cup
1	22	23
2	24	21
3	23	22
4	25	24
5	20	22
6	19	20
7	24	25
8	25	26
9	20	18
10	19	21

At the 0.05 level of significance, is there evidence of a difference in the mean ratings of coffeepot-brewed coffee and K-cup-brewed coffee?

Answers to Problems

1. Because $Z = 3.6579 > 1.96$ (or p-value $= 0.0003 < 0.05$), reject H_0. There is evidence of a difference between men and women in the proportion who say that they prefer the local pizza.

2. Because $Z = 0.6110 < 1.96$ (or p-value $= 0.5412 > 0.05$), do not reject H_0. There is no evidence of a difference in the proportion of females and males who churned.

3. a.   Because $Z = -10.6384 > -1.96$ (or p-value $= 0.00 < 0.05$), reject H_0. There is evidence of a difference in the proportion of customers who churned between customers who used paperless billing and those customers who did not.

   b.   Because $Z = 1.5427 < 1.96$ (or p-value $= 0.1229 < 0.05$), do not reject H_0. There is no evidence of a difference in the proportion of customers who churned between customers who used paperless billing and those customers who did not.

   c. The sample size had a profound effect on the results even though the sample proportions were similar.

4. a. H_0: $\mu_1 = \mu_2$ (The two population means are equal.) The alternative hypothesis is H_1: $\mu_1 \neq \mu_2$ (The two population means are not equal.)

   b. Because $t = -6.1532 < -2.0244$ (or *p*-value = 0.0000 < 0.05), reject H_0. There is evidence of a difference in the mean estimated number of calories in the cheeseburger for the people who thought about the cheesecake first and for those who thought about the organic fruit salad first.

5. Because $t = 2.6411 > 2.0017$ (or *p*-value = 0.0106 < 0.05), reject H_0. There is evidence that a difference exists in the mean time spent per day accessing the Internet via a mobile device between females and males.

6. Because $t = 0.5314 < 1.9954$ (or *p*-value = 0.5968 > 0.05), do not reject H_0. There is insufficient evidence that a difference exists in the mean battery time of a Samsung Galaxy phone and an iPhone.

7. Because $t = -4.5300 < -2.2281$ (or *p*-value = 0.0011 < 0.05), reject H_0. There is evidence of a difference in the mean price at Aldi and Publix.

8. Because $t = 1.8426 < 2.4469$ (or *p*-value = 0.1150 > 0.05), do not reject H_0. There is insufficient evidence of a difference in the mean bone marrow microvessel density before the stem cell transplant and after the stem cell transplant.

9. Because $t = -0.1829 < -2.2622$ (or *p*-value = 0.8589 > 0.01), do not reject H_0. There is insufficient evidence that the mean rating is different for coffeepot-brewed and K-cup-brewed coffees.

References

1. Berenson, M. L., D. M. Levine, K. A. Szabat, and D. Stephan. *Basic Business Statistics: Concepts and Applications*, 15th edition. Hoboken, NJ: Pearson Education, 2023.

2. Levine, D. M., D. Stephan, and K. A. Szabat. *Statistics for Managers Using Microsoft Excel*, 9th edition. Hoboken, NJ: Pearson Education, 2021.

CHAPTER

9

Hypothesis Testing: Chi-Square Tests and the One-Way Analysis of Variance (ANOVA)

In Chapter 8, you learned about several hypothesis tests that you use to analyze differences between two groups. In this chapter, you will learn about tests that you can use when you have multiple (two or more) groups.

9.1   Chi-Square Test for Two-Way Tables

CONCEPT   The hypothesis tests for the difference in the proportion of successes in two or more groups or a relationship between two categorical variables in a two-way table, which Section 2.1 first discusses.

INTERPRETATION   Recall from Chapter 2 that a two-way table presents the count of joint responses to two categorical variables. The categories of one variable form the rows of the table, and the categories of the other variable form the columns. The chi-square test determines whether a relationship exists between the row variable and the column variable.

The null and alternative hypotheses for the two-way table are

H_0: There is no relationship between the row variable and the column variable.

H_1: There is a relationship between the row variable and the column variable.

For the special case of a table that contains only two rows and two columns, the chi-square test becomes equivalent to the Z test for the difference between two proportions that Section 8.1 discusses. The null and alternative hypotheses are restated as

H_0: $\pi_1 = \pi_2$ (No difference exists between the two proportions.)

H_1: $\pi_1 \neq \pi_2$ (A difference exists between the two proportions.)

The chi-square test compares the actual count, or *frequency*, in each cell with the frequency that would be expected to occur if the null hypothesis were true. The expected frequency for each cell is calculated by multiplying the row total of that cell by the column total of that cell and dividing by the total sample size:

$$\text{expected frequency} = \frac{(\text{row total})(\text{column total})}{\text{sample size}}$$

Because some differences are positive and some are negative, each difference is squared and divided by the expected frequency. The results for all cells are then summed to produce a statistic that follows the chi-square distribution.

To use this test, the expected frequency for each cell must be greater than 1, except for the special case of a two-way table that has two rows and two columns, in which the expected frequency of each cell should be at least 5. [If, for this special case, the expected frequency is less than 5, you can use alternative tests, such as Fisher's exact test (see references 2 and 3).]

WORKED-OUT PROBLEM 1   Businesses use a method called *A/B testing* to test different web page designs to see if one design is more effective than another. For one company, designers were interested in the effect of modifying the call-to-action button on the home page.

Every visitor to the company's home page was randomly shown either the original call-to-action button (the control) or the new variation. Designers measured success by the download rate: the number of people who downloaded the file divided by the number of people who saw that particular call-to-action button. Results of the experiment yielded the following:

| | | Call-to-Action Button | | |
		Original	New	Total
Download	Yes	351	451	802
	No	3,291	3,105	6,396
	Total	3,642	3,556	7,198

The row 1 total shows that 802 visitors to the site downloaded the file. The column 1 total shows that 3,642 visitors to the site saw the original call-to-action button. The expected frequency for downloading the new call-to-action button is 396.21, the product of the total number of downloads (802) multiplied by the number of visitors who saw the new call-to-action button (3,556), divided by the total sample size (7,198):

$$\text{expected frequency} = \frac{(802)(3{,}556)}{7{,}198}$$
$$= 396.21$$

The expected frequencies of the four cells are as follows:

| | | Call-to-Action Button | | |
		Original	New	Total
Download	Yes	405.79	396.21	802
	No	3,236.21	3,159.79	6,396
	Total	3,642	3,556	7,198

Figure 9.1 on page 186 presents the spreadsheet results for the chi-square test for this study. The results show that the p-value for this chi-square test is 0.0000. Because the p-value is less than the level of significance, $\alpha = 0.05$, you reject the null hypothesis. You conclude there is evidence that a relationship exists between the call-to-action button and whether the file is downloaded. You assert that a significant difference exists between the original and new call-to-action buttons in the proportion of visitors who download the file.

FIGURE 9.1

Chi-square
test results for
the web page
A/B test study

	A	B	C	D
1	**Chi-Square Test**			
2				
3		**Observed Frequencies**		
4			**Call to Action Button**	
5	**Download?**	Original	New	Total
6	Yes	351	451	802
7	No	3291	3105	6396
8	Total	3642	3556	7198
9				
10		**Expected Frequencies**		
11			**Call to Action Button**	
12	Download?	Original	New	Total
13	Yes	405.7911	396.2089	802
14	No	3236.2089	3159.7911	6396
15	Total	3642	3556	7198
16				
17		**Data**		
18	**Level of Significance**	0.05		
19	Number of Rows	2		
20	Number of Columns	2		
21	Degrees of Freedom	1		
22				
23		**Results**		
24	**Critical Value**	3.8415		
25	**Chi-Square Test Statistic**	16.8527		
26	**p-Value**	0.0000		
27	**Reject the null hypothesis**			
28				
29	*Expected frequency assumption*			
30	*is met.*			

WORKED-OUT PROBLEM 2   You decide to use the critical value approach
for the study concerning the website design. The Figure 9.1 results show that
the chi-square statistic is 16.8527. The number of degrees of freedom for the
chi-square test equals the number of rows minus 1 multiplied by the number of
columns minus 1:

Degrees of freedom = (number of rows − 1) × (number of columns − 1)

Using the table of the chi-square distribution (Table C.3), with $\alpha = 0.05$ and the
degrees of freedom = $(2 − 1)(2 − 1) = 1$, the critical value of chi-square is equal
to 3.841. Because 16.8527 > 3.841, you reject the null hypothesis.

WORKED-OUT PROBLEM 3   Quick-service restaurants are evaluated on
many variables, and the results are summarized periodically in *QSR Magazine*.
One important variable is the accuracy of the order. You seek to determine, with
a level of significance of $\alpha = 0.05$, whether a difference exists in the proportions of food orders filled correctly at the Burger King, Wendy's, and McDonald's
drive-through windows. You use the following data that report the results of
drive-through performance.

		Quick-Service Restaurant		
		Burger King	Wendy's	McDonald's
Order Filled Correctly	**Yes**	909	891	929
	No	91	109	71
	Total	1,000	1,000	1,000

The null and alternative hypotheses are as follows:

H_0: $\pi_1 = \pi_2 = \pi_3$ (No difference exists in the proportion of correct orders among Burger King, Wendy's, and McDonald's.)

H_1: $\pi_1 \neq \pi_2 \neq \pi_3$ (A difference exists in the proportion of correct orders among Burger King, Wendy's, and McDonald's.)

Figure 9.2 presents the spreadsheet results for the chi-square test for this study. Because the p-value for this chi-square test, 0.0123, is less than $\alpha = 0.05$, you reject the null hypothesis. There is evidence of a difference in the proportion of correct orders filled among Burger King, Wendy's, and McDonald's.

FIGURE 9.2

Chi-square test results for the quick-service restaurants study

	A	B	C	D	E
1	Chi-Square Test for Quick-Service Restaurants				
2					
3		Observed Frequencies			
4		Quick-Service Restaurant			
5	Order Filled Correctly	Burger King	McDonald's	Wendy's	Total
6	Yes	909	929	891	2729
7	No	91	71	109	271
8	Total	1000	1000	1000	3000
9					
10		Expected Frequencies			
11		Quick-Service Restaurant			
12	Order Filled Correctly	Burger King	McDonald's	Wendy's	Total
13	Yes	909.6667	909.6667	909.6667	2729
14	No	90.3333	90.3333	90.3333	271
15	Total	1000	1000	1000	3000
16					
17	Data				
18	Level of Significance	0.05			
19	Number of Rows	2			
20	Number of Columns	3			
21	Degrees of Freedom	2			
22					
23	Results				
24	Critical Value	5.9915			
25	Chi-Square Test Statistic	8.7944			
26	p-Value	0.0123			
27	Reject the null hypothesis				
28					
29	Expected frequency assumption				
30	is met.				

WORKED-OUT PROBLEM 4   Using the critical value approach for the same problem, the spreadsheet results show that the chi-square statistic is 8.7944. At the 0.05 level of significance, with the 2 degrees of freedom $[(2 - 1)(3 - 1) = 2]$, the chi-square critical value from Table C.3 is 5.991. Because the chi-square statistic 8.7944 is greater than 5.991, you reject the null hypothesis.

WORKED-OUT PROBLEM 5   A restaurant owner wonders whether there is a relationship, at the 0.05 level of significance, between the type of dessert and type of entrée ordered by guests during the Friday-to-Sunday weekend period. The owner records the desserts and entrées ordered by 630 guests during a Friday-to-Sunday weekend period and summarizes the data collection in this two-way table:

		Type of Entrée				
		Beef	**Poultry**	**Fish**	**Pasta**	**Total**
	Ice Cream	13	8	12	14	47
	Cake	98	12	29	6	145
Type of Dessert	**Fruit**	8	10	6	2	26
	None	124	98	149	41	412
	Total	243	128	196	63	630

The null and alternative hypotheses are

H_0: No relationship exists between the type of dessert ordered and the type of entrée ordered.

H_1: A relationship exists between the type of dessert ordered and the type of entrée ordered.

Figure 9.3 presents the spreadsheet results for the chi-square test for this problem. Because the p-value for this chi-square test, 0.0000, is less than $\alpha = 0.05$, you reject the null hypothesis. Evidence exists of a relationship between the type of dessert ordered and the type of entrée ordered.

WORKED-OUT PROBLEM 6   Using the critical value approach for the same problem, the Figure 9.3 results on page 189 show that the chi-square statistic is 92.1028. At the 0.05 level of significance with the 9 degrees of freedom $[(4 - 1)(4 - 1) = 9]$, the chi-square critical value from Table C.3 is 16.919. Because the chi-square test statistic of 92.1028 is greater than 16.919, you reject the null hypothesis.

FIGURE 9.3

Chi-square test results for the type of entrée and dessert problem

	A	B	C	D	E	F
1	Chi-Square Test for Restaurant Study					
2						
3		Observed Frequencies				
4			Type of Entrée			
5	Type of Dessert	Beef	Poultry	Fish	Pasta	Total
6	Ice cream	13	8	12	14	47
7	Cake	98	12	29	6	145
8	Fruit	8	10	6	2	26
9	None	124	98	149	41	412
10	Total	243	128	196	63	630
11						
12		Expected Frequencies				
13			Type of Entrée			
14	Type of Dessert	Beef	Poultry	Fish	Pasta	Total
15	Ice cream	18.1286	9.5492	14.6222	4.7000	47
16	Cake	55.9286	29.4603	45.1111	14.5000	145
17	Fruit	10.0286	5.2825	8.0889	2.6000	26
18	None	158.9143	83.7079	128.1778	41.2000	412
19	Total	243	128	196	63	630
20						
21	Data					
22	Level of Significance	0.05				
23	Number of Rows	4				
24	Number of Columns	4				
25	Degrees of Freedom	9				
26						
27	Results					
28	Critical Value	16.9190				
29	Chi-Square Test Statistic	92.1028				
30	p-Value	0.0000				
31	Reject the null hypothesis					
32						
33	Expected frequency assumption					
34	is met.					

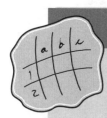

spreadsheet solution

Chi-Square Tests

Chapter 9 Chi-Square contains the spreadsheet that WORKED-OUT PROBLEMS 1 and 2 use to perform the chi-square test.

Chapter 9 Chi-Square Spreadsheets contains the spreadsheet that WORKED-OUT PROBLEMS 3 through 6 use to perform other chi-square tests that have more than two rows and two columns.

Best Practices

Use the **CHISQ.INV.RT**(*level of significance, degrees of freedom*) function to calculate a critical value of the chi-square distribution.

Use the **CHISQ.DIST.RT**(*critical value, degrees of freedom*) function to calculate a probability associated with the chi-square distribution.

How-Tos

For other problems, change the contents of the observed frequencies table that begins in row 3.

Experiment by changing the level of significance (cell B18 in the Chapter 9 Chi-Square spreadsheet).

Chapter 9 Chi-Square Spreadsheets contains these spreadsheets: ChiSquare2×2 (the spreadsheet in Chapter 9 Chi-Square), ChiSquare2×3, ChiSquare3×4, and ChiSquare4×4. To use a specific spreadsheet, click on the sheet tab for the spreadsheet.

equation blackboard (optional)

interested in math?

You need the subscripted symbols for the **observed cell frequencies**, f_0, and the **expected cell frequencies**, f_e, to write the equation for the chi-square test for a two-way table:

$$\chi^2 = \sum_{all\ cells} \frac{\left(f_0 - f_e\right)^2}{f_e}$$

For WORKED-OUT PROBLEM 1, the web page A/B test study, the calculations are:

f_0	f_t	$(f_0 - f_t)$	$(f_0 - f_t)^2$	$(f_0 - f_t)^2/f_t$
351	405.7911	−54.7911	3,002.0595	7.3980
3,291	3,236.2089	54.7911	3,002.0595	0.9276
451	396.2089	54.7911	3,002.0595	7.5770
3,105	3,159.7911	−54.7911	3,002.0595	0.9501
				16.8527

Using the level of significance $\alpha = 0.05$, with $(2 - 1)(2 - 1) = 1$ degree of freedom, from Table C.3, the critical value is 3.841. Because $16.85273 > 3.841$, you reject the null hypothesis.

9.2 One-Way Analysis of Variance (ANOVA): Testing for the Differences Among the Means of More Than Two Groups

Many analyses involve experiments that test whether differences exist in the means of more than two groups. Evaluating differences between groups can be considered a one-factor experiment, also known as a *completely randomized design*.

Factor

CONCEPT   The variable that defines the groups being used for tests of differences in the means among multiple groups. The values for the variable serving as a factor are called *levels*, and a set of levels may be either a set of discrete numerical values or a set of categories.

EXAMPLE   The variable baking temperature with the numerical levels 300°, 350°, 400°, and 450° for an industrial process study; the variable learning materials used with the categorical levels Set A, Set B, and Set C for an educational research study.

One-Way ANOVA

CONCEPT   The hypothesis test that simultaneously compares the differences among the population means of more than two groups in a one-factor experiment.

INTERPRETATION   Unlike the *t* test, which compares differences in two means, the analysis of variance simultaneously compares the differences among the means of more than two groups. Although ANOVA is an acronym for **AN**alysis **O**f **VA**riance, the term is misleading because the objective in the analysis of variance is to analyze differences among the group means, *not* the variances. The null and alternative hypotheses are

H_0: All the population means are equal.

H_1: Not all the population means are equal.

In ANOVA, the total variation in the values is subdivided into variation that is due to differences among the groups and variation that is due to variation within the groups, as Figure 9.4 on page 192 illustrates.

FIGURE 9.4

Subdividing
the total
variation (SST)

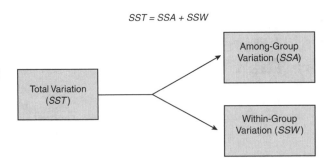

The within-group variation is the **experimental error**, and the among-group variation, which represents the variation due to the factor of interest, is the **treatment effect**. The **sum of squares total (SST)** is the total variation that represents the sum of the squared differences between each individual value and the mean of all the values:

$$SST = \text{The sum of (Each value} - \text{Mean of all values)}^2$$

The among-group variation is known as the **sum of squares among groups (SSA)** and is the sum of the squared differences between the sample mean of each group and the mean of all the values, weighted by the sample size in each group:

$$SSA = \text{The sum of [(Group sample size)} \times \text{(Group mean} - \text{Mean of all values)}^2]$$

The within-group variation is known as the **sum of squares within groups (SSW)** and is the sum of the squared difference between each value and the mean of its own group for all groups:

$$SSW = \text{The sum of (Each value in the group} - \text{Group mean)}^2$$

The Three Variances of ANOVA

ANOVA derives its name from the fact that the differences between the means of the groups are analyzed by comparing variances. Section 3.3 defines the variance as the sum of squared differences around the mean divided by the sample size minus 1:

$$\text{Variance} = S^2 = \frac{\text{sum of squared differences around the mean}}{\text{sample size} - 1}$$

This sample size minus 1 is the **degrees of freedom**, the actual number of values that are free to vary once the mean is known.

In the analysis of variance, there are three different variances: the variance among groups, the variance within groups, and the total variance. Collectively, these variances are called the **mean squares**.

The **mean square among groups (MSA)** is equal to the sum of squares among groups (*SSA*) divided by the number of groups minus 1. The **mean square within groups (MSW)** is equal to the sum of squares within groups (*SSW*) divided by the sample size minus the number of groups. The **mean square total (MST)** is equal to the sum of squares total (*SST*) divided by the sample size minus 1.

To test the null hypothesis

H_0: All the population means are equal.

against the alternative hypothesis

H_1: Not all the population means are equal.

you calculate the test statistic *F*, which follows the *F* distribution (see Table C.4), as the ratio of two of the variances, *MSA* to *MSW*:

$$F = \frac{MSA}{MSW}$$

ANOVA Summary Table

Table 9.1 shows the typical way that analysis of variance results are presented in a summary table. Such a table includes the three sources of variation (among-group, within-group, and total), the degrees of freedom, the sums of squares, the mean squares (or variances), and the *F* test statistic. When software procedures the summary table, a *p*-value may be included in the table.

TABLE 9.1   Generalized ANOVA summary table

Source	Degrees of Freedom	Sum of Squares	Mean Square (Variance)	F
Among groups	number of groups – 1	SSA	$MSA = \dfrac{SSA}{\text{number of groups} - 1}$	$F = \dfrac{MSA}{MSW}$
Within groups	sample size – number of groups	SSW	$MSW = \dfrac{SSW}{\text{sample size} - \text{number of groups}}$	
Total	sample size – 1	SST		

After performing a one-way ANOVA and finding a significant difference among groups, you do not know which groups are significantly different. All that is known is that sufficient evidence exists to state that the population means are not all the same. To determine exactly which groups differ, all possible pairs of groups need to be compared. Many statistical procedures for making these comparisons have been developed (see references 1, 5, 6, 7).

WORKED-OUT PROBLEM 7   You want to determine, with a level of significance $\alpha = 0.05$, whether differences exist among three sets of mathematics learning materials (labeled A, B, and C). You devise an experiment that randomly assigns 24 students to one of the three sets of materials. At the end of a school year, all 24 students are given the same standardized mathematics test that is scored on a 0 to 100 scale. The results of that test are as follows:

Math

Set A	Set B	Set C
87	58	81
80	63	62
74	64	70
82	75	64
74	70	70
81	73	72
97	80	92
71	62	63

Figure 9.5 presents the spreadsheet results for the analysis of variance for the learning materials analysis. Because the p-value for this test, 0.0259, is less than $\alpha = 0.05$, you reject the null hypothesis. You conclude that the mean scores are not the same for all the sets of mathematics materials. From the results, you see that the mean for materials A is 80.75, for materials B it is 68.125, and for materials C it is 71.75. This suggests that the mean score is higher for materials A than for materials B and C.

FIGURE 9.5

One-way ANOVA results for the learning materials analysis

	A	B	C	D	E	F	G	H	I	J
1	One-Way ANOVA for Learning Materials Study									Calculations
2									c	3
3	SUMMARY								n	24
4	Groups	Count	Sum	Average	Variance					
5	Set A	8	646	80.75	70.21429					
6	Set B	8	545	68.125	56.98214					
7	Set C	8	574	71.75	104.7857					
8										
9										
10	ANOVA									
11	Source of Variation	SS	df	MS	F	P-value	F crit			
12	Between Groups	676.0833	2	338.0417	4.3716	0.0259	3.4668			
13	Within Groups	1623.8750	21	77.3274						
14										
15	Total	2299.9583	23							
16					Level of significance	0.05				

WORKED-OUT PROBLEM 8   You want to use the critical value approach for the same problem. This approach requires you to look up the critical value of F in a table of critical values of F such as Table C.4. To perform a lookup, you use these degrees of freedom:

- The numerator degrees of freedom, equal to the number of groups minus 1
- The denominator degrees of freedom, equal to the sample size minus the number of groups

With three groups and a sample size of 24, the numerator degrees of freedom are 2 ($3 - 1 = 2$), and the denominator degrees of freedom are 21 ($24 - 3 = 21$). From Table C.4, given a level of significance $\alpha = 0.05$, the critical value of F is 3.47. Because the decision rule is to reject H_0 if $F >$ critical value of F, and $F = 4.37 > 3.47$, you reject the null hypothesis.

The critical value of F also appears in the one-way ANOVA spreadsheet results in this book. In the Figure 9.5 results, the critical value appears in cell G12 as the value 3.4668, which rounds to 3.47.

equation
blackboard
(optional)

interested
in
math?

To form the equations for the three mean squares and the test statistic F, you assemble these symbols:

- $\overline{\overline{X}}$, pronounced as "X double bar," that represents the overall or grand mean

- a subscripted X Bar, $\overline{X}_j$, that represents the mean of a group

- a double-subscripted uppercase italic X, X_{ij}, that represents individual values in group j

- a subscripted lowercase italic n, n_j, that represents the sample size in a group

- a lowercase italic n, n, that represents the total sample size (the sum of the sample sizes of each group)

- a lowercase italic c, c, that represents the number of groups

First, you form the equation for the grand mean as

$$\overline{\overline{X}} = \frac{\displaystyle\sum_{j=1}^{c}\sum_{i=1}^{n_j} X_{ij}}{n} = \text{grand mean}$$

$\overline{X}_j$ = sample mean of group j

X_{ij} = ith value in group j

n_j = number of values in group j

n = total number of values in all groups combined

(that is, $n = n_1 + n_2 + \cdots + n_c$)

c = number of groups of the factor of interest

With $\overline{\overline{X}}$ defined, you then form the equations that define the sum of squares total, SST, the sum of squares among groups, SSA, and the sum of squares within groups, SSW:

$$SST = \sum_{j=1}^{c}\sum_{i=1}^{n_j} (X_{ij} - \overline{\overline{X}})^2$$

$$SSA = \sum_{j=1}^{c} n_j (X_j - \overline{\overline{X}})^2$$

$$SSW = \sum_{j=1}^{c}\sum_{i=1}^{n_j} (X_{ij} - \overline{X}_j)^2$$

Next, using these definitions, you form the equations for the mean squares:

$$MSA = \frac{SSA}{\text{number of groups} - 1}$$

$$MSW = \frac{SSW}{\text{sample size} - \text{number of groups}}$$

$$MST = \frac{SST}{\text{sample size} - 1}$$

Finally, using the definitions of MSA and MSW, you form the equation for the test statistic F:

$$F = \frac{MSA}{MSW}$$

Using the data from WORKED-OUT PROBLEM 7,

$$\overline{\overline{X}} = \frac{1,765}{24} = 73.5417$$

$$SST = \sum_{j=1}^{c}\sum_{i=1}^{n_j}(X_{ij} - \overline{\overline{X}})^2 = (87 - 73.5417)^2 + \cdots + (71 - 73.5417)^2$$

$$+ (58 - 73.5417)^2 + \cdots + (62 - 73.5417)^2$$

$$+ (81 - 73.5417)^2 + \cdots + (63 - 73.5417)^2 = 2,299.9583$$

$$SSA = \sum_{j=1}^{c} n_j (X_{ij} - \overline{\overline{X}})^2 = 8(80.75 - 73.5417)^2$$

$$+ 8(68.125 - 73.5417)^2 + 8(71.75 - 73.5417)^2$$

$$= 676.0833$$

$$SSW = \sum_{j=1}^{c}\sum_{i=1}^{n_j}(X_{ij} - \overline{X}_j)^2 = (87 - 80.75)^2 + \cdots + (71 - 80.75)^2$$

$$+ (58 - 68.125)^2 + \cdots + (62 - 68.125)^2$$

$$+ (81 - 71.75)^2 + \cdots + (63 - 71.75)^2 = 1,623.875$$

Using the previous calculations:

$$MSA = \frac{SSA}{\text{number of groups} - 1} = \frac{676.0833}{2} = 338.0417$$

$$MSW = \frac{SSW}{\text{sample size} - \text{number of groups}} = \frac{1,623.8750}{21} = 77.3274$$

$$F = \frac{MSA}{MSW} = \frac{338.0417}{77.3274} = 4.3716$$

Because $F = 4.3716 > 3.47$, you reject H_0.

spreadsheet solution

Chapter 9 One-Way ANOVA contains the Figure 9.5 spreadsheet that WORKED-OUT PROBLEM 7 uses to perform a one-way ANOVA.

Best Practices

Use the **COUNT, SUM, AVERAGE**, and **VAR.S** functions to calculate the counts, sums, averages, and variances, respectively.

Use the **F.INV.RT**(*level of significance*, *degrees of freedom between groups*, *degrees of freedom within groups*) function to calculate the critical value of the F distribution.

Use the **F.DIST.RT**(F *test statistic*, *degrees of freedom between groups*, *degrees of freedom within groups*) function to calculate the *p*-value.

Use the **DEVSQ**(*cell range of the one-way ANOVA data*) function to compute the sum of the squares for the total variation (*SST*).

How-Tos

Experiment with this spreadsheet by changing the level of significance in cell G16.

Advanced Technique ATT5 in Appendix E describes using the Analysis ToolPak as a second way to perform a one-way ANOVA.

Modifying **Chapter 9 One-Way ANOVA** for use with other problems is beyond the scope of this book. However, **Chapter 9 WORKED-OUT PROBLEM 9** presents the modifications needed in the spreadsheet SUMMARY and ANOVA tables to apply the spreadsheet solution to WORKED-OUT PROBLEM 9.

WORKED-OUT PROBLEM 9   A pet food company is looking to expand its product line beyond its current kidney- and shrimp-based cat foods. The company developed two new products: one based on chicken livers and the other based on salmon. The company conducted an experiment to compare the two new products with its two existing ones as well as a generic beef-based product sold by a supermarket chain.

For the experiment, a sample of 50 cats from the population at a local animal shelter was selected. Ten cats were randomly assigned to each of the five products being tested. Each of the cats was then presented with 3 ounces of the selected food in a dish at feeding time. The researchers defined the variable to be measured as the number of ounces of food that the cat consumed within a 10-minute time interval that began when the filled dish was presented. The results for this experiment are summarized in the following table.

Cat Food

Kidney	Shrimp	Chicken Liver	Salmon	Beef
2.37	2.26	2.29	1.79	2.09
2.62	2.69	2.23	2.33	1.87
2.31	2.25	2.41	1.96	1.67
2.47	2.45	2.68	2.05	1.64
2.59	2.34	2.25	2.26	2.16
2.62	2.37	2.17	2.24	1.75
2.34	2.22	2.37	1.96	1.18
2.47	2.56	2.26	1.58	1.92
2.45	2.36	2.45	2.18	1.32
2.32	2.59	2.57	1.93	1.94

Figure 9.6 presents the spreadsheet results for the analysis of variance for the cat food study. Because the p-value for this test, 0.000, is less than $\alpha = 0.05$, you reject the null hypothesis. You conclude that evidence of a difference exists in the mean amount of food eaten among the five types of cat foods.

FIGURE 9.6

One-way ANOVA results for the cat food study

	A	B	C	D	E	F	G	H	I	J
1	One-Way ANOVA for Cat Food Study								Calculations	
2									c	5
3	SUMMARY								n	50
4	Groups	Count	Sum	Average	Variance					
5	Kidney	10	24.56	2.456	0.0148					
6	Shrimp	10	24.09	2.409	0.0253					
7	Chicken Liver	10	23.68	2.368	0.0263					
8	Salmon	10	20.28	2.028	0.0544					
9	Beef	10	17.54	1.754	0.0990					
10										
11										
12	ANOVA									
13	Source of Variation	SS	df	MS	F	P-value	F crit			
14	Between Groups	3.6590	4	0.9147	20.8054	0.0000	2.5787			
15	Within Groups	1.9785	45	0.0440						
16										
17	Total	5.6375	49							
18						Level of significance	0.05			

WORKED-OUT PROBLEM 10   You want to use the critical value approach for the same problem. Using the Figure 9.6 spreadsheet results in lieu of Table C.4, the critical value of F is 2.5787 (cell G14), given a level of significance $\alpha = 0.05$, with 4 degrees of freedom in the numerator (5 − 1) and 45 degrees of freedom in the denominator $(50 - 5)$. The computed F test statistic for this problem is 20.8054 (cell E14). Because the F test statistic 20.8054 is greater than 2.5787, you reject the null hypothesis.

One-Way ANOVA Assumptions

To use the one-way ANOVA F test, you must make three major assumptions: randomness and independence, normality, and homogeneity of variance.

The first assumption, randomness and independence, always must be met, because the validity of your experiment depends on the random sampling or random assignment of items or subjects to groups. Departures from this assumption can seriously affect inferences from the analysis of variance. (References 6 and 7 provide more discussion of these issues.)

The second assumption, normality, states that the values in each group are selected from normally distributed populations. The one-way ANOVA F test is not very sensitive to departures from this assumption of normality. As long as the distributions are not very skewed, the level of significance of the ANOVA F test is usually not greatly affected by lack of normality, particularly for large samples. When only the normality assumption is seriously violated, nonparametric alternatives to the one-way ANOVA F test are available (see references 1 and 5).

The third assumption, equality of variances, states that the variance within a population should be equal for all populations. Although the one-way ANOVA F test is relatively robust or insensitive with respect to the assumption of equal group variances, large departures from this assumption can seriously affect the level of significance and the power of the test. Therefore, various procedures have been developed to test the assumption of homogeneity of variance (see references 1, 4, and 5).

One way to evaluate the assumptions is to construct side-by-side boxplots of the groups to study their central tendency, variation, and shape.

Other Experimental Designs

The one-way analysis of variance is the simplest type of experimental design because it considers only one factor of interest. More complicated experimental designs examine at least two factors of interest simultaneously. For more information on these designs, see references 1, 4, 6, and 7.

Important Equations

Chi-Square Test for a Two-Way Table:

$$\chi^2 = \sum_{all\ cells} \frac{(f_0 - f_e)^2}{f_e}$$

ANOVA Calculations:

$$SST = \sum_{j=1}^{c} \sum_{i=1}^{n_j} (X_{ij} - \overline{\overline{X}})^2$$

$$SSA = \sum_{j=1}^{c} n_j (X_j - \overline{\overline{X}})^2$$

$$SSW = \sum_{j=1}^{c} \sum_{i=1}^{n_j} (X_{ij} - \overline{X}_j)^2$$

$$MSA = \frac{SSA}{number\ of\ groups - 1}$$

$$MSW = \frac{SSW}{sample\ size - number\ of\ groups}$$

$$MST = \frac{SST}{sample\ size - 1}$$

$$F = \frac{MSA}{MSW}$$

One-Minute Summary

Tests for the differences among more than two groups:

- If your data are categorical, use chi-square (χ^2) tests (which can also be used for two groups).
- If your data are numerical and if you have one factor, use the one-way ANOVA.

Test Yourself
Short Answers

1. In a one-way ANOVA, if the F test statistic is greater than the critical F value, you:

    a. reject H_0 because there is evidence all the means differ

    b. reject H_0 because there is evidence at least one of the means differs from the others

    c. do not reject H_0 because there is no evidence of a difference in the means

    d. do not reject H_0 because one mean is different from the others

2. In a one-way ANOVA, if the p-value is greater than the level of significance, you:

    a. reject H_0 because there is evidence all the means differ

    b. reject H_0 because there is evidence at least one of the means differs from the others

    c. do not reject H_0 because there is insufficient evidence of a difference in the means

    d. do not reject H_0 because one mean is different from the others

3. The F test statistic in a one-way ANOVA is:

    a. MSW/MSA

    b. SSW/SSA

    c. MSA/MSW

    d. SSA/SSW

4. In a one-way ANOVA, the null hypothesis is always:

    a. all the population means are different

    b. some of the population means are different

    c. some of the population means are the same

    d. all of the population means are the same

5. A car rental company wants to select a computer software package for its reservation system. Three software packages (A, B, and C) are commercially available. The car rental company will choose the package that has the lowest mean number of renters for whom a car is not available at the time of pickup. An experiment is set up in which each package is used to make reservations for five randomly selected weeks. How should the data be analyzed?

    a. chi-square test for differences in proportions

    b. one-way ANOVA F test

    c. t test for the differences in means

    d. t test for the mean difference

The following information should be used to answer Questions 6 through 9:

For quick-service restaurants, the drive-through window is an increasing source of revenue. The chain that offers that fastest service is considered most likely to attract additional customers. For a study of 20 drive-through times (from menu board to departure) at 5 quick-service chains, the following ANOVA table was developed.

Source	DF	Sum of Squares	Mean Squares	F
Among Groups (Chains)		6,536	1,634.0	12.51
Within Groups (Chains)	95		130.6	
Total	99	18,943		

6. Referring to the preceding table, the among groups degrees of freedom is:
   a. 3
   b. 4
   c. 12
   d. 16

7. Referring to the preceding table, the within groups sum of squares is:
   a. 12,407
   b. 95
   c. 130.6
   d. 4

8. Referring to the preceding table, the within groups mean square is:
   a. 12,407
   b. 95
   c. 130.6
   d. 4

9. Referring to the preceding table, at the 0.05 level of significance, you:
   a. do not reject the null hypothesis and conclude that no difference exists in the mean drive-up time between the chains
   b. do not reject the null hypothesis and conclude that a difference exists in the mean drive-up time between the chains
   c. reject the null hypothesis and conclude that a difference exists in the mean drive-up time between the chains
   d. reject the null hypothesis and conclude that no difference exists in the mean drive-up time between the chains

10. When testing for independence in a contingency table with three rows and four columns, there are _____ degrees of freedom.

    a.   5

    b.   6

    c.   7

    d.   12

11. In testing a hypothesis using the chi-square test, the theoretical frequencies are based on the:

    a.   null hypothesis

    b.   alternative hypothesis

    c.   normal distribution

    d. t distribution

12. An agronomist is studying three different varieties of tomato to determine whether a difference exists in the proportion of seeds that germinate. Random samples of 100 seeds of each of three varieties are subjected to the same starting conditions. How should the data be analyzed?

    a.   chi-square test for differences in proportions

    b.   one-way ANOVA F test

    c. t test for the differences in means

    d. t test for the mean difference

Answer True or False:

13. A test for the difference between two proportions can be performed using the chi-square distribution.

14. The one-way analysis of variance (ANOVA) tests hypotheses about the difference between population proportions.

15. The one-way analysis of variance (ANOVA) tests hypotheses about the difference between population means.

16. The one-way analysis of variance (ANOVA) tests hypotheses about the difference between population variances.

17. The mean squares in an ANOVA can never be negative.

18. In a one-factor ANOVA, the among sum of squares and within sum of squares must add up to the total sum of squares.

19. If you use the chi-square method of analysis to test for the difference between two proportions, you must assume that there are at least five observed frequencies in each cell of the contingency table.

20. If you use the chi-square method of analysis to test for independence in a contingency table with more than two rows and more than two columns, you must assume that there is at least one theoretical frequency in each cell of the contingency table.

Answers to Test Yourself Short Answers

1. b	11. a
2. c	12. a
3. c	13. True
4. d	14. False
5. b	15. True
6. b	16. False
7. a	17. True
8. c	18. True
9. c	19. True
10. b	20. True

Problems

1. Students who attend a regional university located in a small town are known to favor the local pizza restaurant. A national chain of pizza restaurants looks to conduct a survey of students to determine pizza preferences. The chain selects a sample of 220 students attending the university. Of 93 females sampled, 74 preferred the local pizza restaurant. Of 127 males sampled, 71 preferred the local pizza restaurant.

    a. At the 0.05 level of significance, is there evidence of a difference in the proportion of students who prefer the local pizza restaurant between females and males?

    b. Compare the part (a) results with those of Problem 1 in Chapter 8.

2. Churning, the loss of customers to a competitor, is especially a problem for telecommunications companies. Market researchers at a telecommunications company collect data from a sample of 5,517 customers. Of 2,741 females selected, 883 churned in the past month. Of 2,776 males selected, 873 churned in the past month.

    a. At the 0.05 level of significance, is there evidence of a difference in the proportion of customers who churned between females and males?

    b. Compare the results of part (a) with those of Problem 2 in Chapter 8.

3. Market researchers at the company in Problem 2 reviewed the same sample, looking for a pattern between churning and paperless billing. Of the 3,725 customers who used paperless billing, 1,358 churned in the past month. Of the 1,792 customers who did not use paperless billing, 398 churned in the past month.

    a. At the 0.05 level of significance, is there evidence of a difference in the proportion of customers who churned between customers who used paperless billing and customers who did not use paperless billing?

    b. Compare the results of part (a) with those of Problem 3 in Chapter 8.

4. You seek to determine, with a level of significance of $\alpha = 0.05$, whether there was a relationship between numbers selected for the Vietnam War–era military draft lottery system and the time of year a man was born. The following shows how many low (1–122), medium (123–244), and high (245–366) numbers were drawn for birth dates in each quarter of the year.

		Quarter of Year				
		Jan–Mar	Apr–Jun	Jul–Sep	Oct–Dec	Total
	Low	21	28	35	38	122
Number Set	Medium	34	22	29	37	122
	High	36	41	28	17	122
	Total	91	91	92	92	366

At the 0.05 level of significance, is there a relationship between draft number and quarter of the year?

5. You want to determine, with a level of significance $\alpha = 0.05$, whether differences exist among the four plants that fill boxes of a particular brand of cereal. You select samples of 20 cereal boxes from each of the four plants. The weights of these cereal boxes (in grams) are as follows.

Box Fills

Plant 1		Plant 2		Plant 3		Plant 4	
361.43	364.78	370.26	360.27	367.53	390.12	361.95	369.36
368.91	376.75	357.19	362.54	388.36	335.27	381.95	363.11
365.78	353.37	360.64	352.22	359.33	366.37	383.90	400.18
389.70	372.73	398.68	347.28	367.60	371.49	358.07	358.61
390.96	363.91	380.86	350.43	358.06	358.01	382.40	370.87
372.62	375.68	334.95	376.50	369.93	373.18	386.20	380.56
390.69	380.98	359.26	369.27	355.84	377.40	373.47	376.21
364.93	354.61	389.56	377.36	382.08	396.30	381.16	380.97
387.13	378.03	371.38	368.50	381.45	354.82	379.41	365.78
360.77	374.24	373.06	363.86	356.20	383.78	382.01	395.55

What conclusions can you reach?

6. *QSR* reports on the largest quick-serve and fast-casual restaurants in the United States. Do the various market segments (burger, chicken, sandwich, and pizza) differ in their mean sales per unit? The following table presents the mean sales per unit by segment in a recent year.

QSR Sales

Burger	Chicken	Sandwich	Pizza
2,788.73	4,166.67	420.00	1,123.00
1,636.40	1,210.00	2,739.91	878.00
1,367.00	1,325.00	1,177.00	874.00
1,250.00	2,060.00	772.00	815.00
1,553.00	1,741.08	769.00	568.00
2,828.00	1,138.00	711.66	721.99
1,131.00	2,963.00	1,027.00	
1,274.80	1,798.00	1,633.51	
2,390.00	714.00	2,520.00	
1,738.00	1,259.00		
2,794.00			
972.42			
1,480.00			

Source: Data extracted from "Ranking The Top 50 Fast-Food Chains in America," https://bit.ly/3mwEjl2.

At the 0.05 level of significance, is there evidence of a difference in the mean sales per unit among the market segments?

7. Managers for a convenience store chain seek to determine the best location for selling small flashlights. They cannot decide whether to sell the flashlights at the checkout counter or in an end-aisle or in-aisle display. They decide to conduct an experiment using a sample of 24 stores that have customer counts that are nearly identical to the mean per-store count for the chain.

In 8 of the stores, the flashlights will be placed at the checkout counter. In 8 of the stores, the flashlights will be placed in an end-aisle display. In 8 of the stores, the flashlights will be placed in an in-aisle display. After four weeks, the managers end their experiment and compile this summary table:

Flashlight Sales

Checkout	End-Aisle	In-Aisle
62	55	44
57	47	37
39	63	28
46	32	59
75	55	41
99	49	32
84	37	50
56	60	29

At the 0.05 level of significance, is there evidence of a difference in the mean sales of flashlights among the three sales locations?

Answers to Problems

1. a.   Because the *p*-value for this chi-square test, 0.0003, is less than $\alpha = 0.05$ (or the chi-square statistic = 13.3804 > 3.841), you reject the null hypothesis. There is evidence of a difference between men and women in the proportion who prefer the local restaurant.

   b.   The results are the same because the chi-square statistic with 1 degree of freedom is the square of the Z statistic.

2. a.   Because the *p*-value for this chi-square test, 0.5412, is greater than $\alpha = 0.05$ (or the chi-square statistic = 0.3733 < 3.841), you do not reject the null hypothesis. There is insufficient evidence of a difference in the proportion of males and the proportion of females who churned.

   b.   The results are the same because the chi-square statistic with 1 degree of freedom is the square of the Z statistic.

3. a.   Because the *p*-value for this chi-square test, 0.0000, is less than $\alpha = 0.05$ (or the chi-square statistic = 113.1772 > 3.841), you reject the null hypothesis. There is evidence of a difference in the proportion who churned between those who used paperless billing and those who did not.

   b.   The results are the same because the chi-square statistic with one degree of freedom is the square of the Z statistic.

4. Because the *p*-value for this chi-square test, 0.0021, is less than $\alpha = 0.05$ (or the chi-square statistic = 20.6804 > 12.5916), you reject the null hypothesis. Evidence exists of a relationship between the number selected and the time of the year when a man was born. It appears that men who were born between January and June were more likely than expected to have high numbers, whereas men born between July and December were more likely than expected to have low numbers.

5. Because the *p*-value for this test, 0.0959, is greater than $\alpha = 0.05$ (or the F test statistic, 2.1913, is less than the critical value of F = 2.725), you cannot reject the null hypothesis. You conclude that there is insufficient evidence of a difference in the mean cereal weights among the four plants.

6. Because the *p*-value for this test, 0.0518, is greater than $\alpha = 0.05$ (or the F test statistic, 2.8505, is less than the critical value of F = 2.8826), you cannot reject the null hypothesis. You conclude that there is insufficient evidence of a difference in the mean sales per unit among the market segments.

7. Because the p-value $= 0.0094 < 0.05$ (or $F = 5.8748 > 3.4668$), reject H_0. There is evidence of a difference among the mean sales of flashlights in the three locations.

References

1. Berenson, M. L., D. M. Levine, K. A. Szabat, and D. Stephan. *Basic Business Statistics: Concepts and Applications*, 15th edition. Hoboken, NJ: Pearson Education, 2023.

2. Conover, W. J. *Practical Nonparametric Statistics*, 3rd edition. New York: Wiley, 2000.

3. Daniel, W. *Applied Nonparametric Statistics*, 2nd edition. Boston: Houghton Mifflin, 1990.

4. Levine, D. M. *Statistics for Six Sigma Green Belts Using Minitab and JMP*. Upper Saddle River, NJ: Prentice Hall, 2006.

5. Levine, D. M., D. Stephan, and K. A. Szabat. *Statistics for Managers Using Microsoft Excel*, 9th edition. Hoboken, NJ: Pearson Education, 2021.

6. Montgomery, D. C. *Design and Analysis of Experiments*, 6th edition. New York: John Wiley, 2005.

7. Kutner, M. H., C. Nachtsheim, J. Neter, and W. Li. *Applied Linear Statistical Models*, 5th edition, New York: McGraw-Hill-Irwin, 2005.

Simple Linear Regression

Regression methods are inferential methods that enable you to use the values of one variable to predict the values of another variable. For example, managers of a growing chain of retail stores might wonder if larger-sized stores generate greater sales, farmers might want to predict the weight of a pumpkin based on its circumference, and baseball fans might want to predict the number of games a team will win in a season based on the number of runs the team scores.

In this chapter, you will learn the basics of regression analysis and specifics about *simple linear regression models* that examine the relationship between two numerical variables.

10.1 Basics of Regression Analysis

Learning regression analysis begins with understanding several new concepts and vocabulary terms that form the basis for regression analysis.

Prediction

CONCEPT   The term used to describe what a regression model does.

INTERPRETATION   Prediction, in the statistical sense, explains how the values of one or more variables are related to another variable. That explanation takes the form of a mathematical expression.

For example, for a retail store chain, it might be that increasing the size of an individual store by one-third increases sales at that store by 50% for stores ranging in size from 3,000 to 7,000 square feet. If a 3,000-square-foot store had sales of $10,000 per week, you could *predict* that increasing the size of the store to 4,000 square feet would likely result in sales of about $15,000 per week.

Dependent Variable

CONCEPT   The variable that a regression model seeks to predict or estimate. The dependent variable is also known as the response variable, the predicted variable, or the *Y* variable.

EXAMPLE   The dependent variable store sales in the example given in the Interpretation section for "Prediction."

Independent Variable

CONCEPT   A variable that a regression model seeks to use to predict the dependent variable. The independent variable is also known as the explanatory variable, the predictor variable, or the *X* variable.

EXAMPLE   The independent variable store size being used to predict the dependent variable store sales in the example given in the Interpretation section for "Prediction."

INTERPRETATION   Statistical prediction is not psychic guessing, nor is it a statement about future events. When choosing an independent variable, there must be a logical relationship between the independent variable and the dependent variable.

Many relationships among logically unrelated variables happen by chance. For example, a researcher once showed how increasing the imports of Mexican lemons decreased U.S. highway fatalities: For every increase of about 300 metric tons of lemons imported, there was a drop of 1 fatality per 100,000 people in the rate of fatalities (see reference 4). Such a non-logical relationship is called a *spurious correlation*.

Scatter Plot

CONCEPT A scatter plot visualizes the relationship in the special case of one independent variable and one dependent variable.

EXAMPLE The following Interpretation section presents several examples of scatter plots.

INTERPRETATION A scatter plot can suggest a specific type of relationship between the X and Y variables that regression analysis can later explore. The Figure 10.1 scatter plots visualize six different kinds of patterns. These patterns are:

- **Panel A:** Positive straight-line relationship between X and Y. As the value of the X variable increases, the value of the Y variable increases.
- **Panel B:** Negative straight-line relationship between X and Y.
- **Panel C:** Positive curvilinear relationship between X and Y. The values of Y are increasing as X increases, but this increase tapers off beyond certain values of X.
- **Panel D:** U-shaped relationship between X and Y. As X increases, at first Y decreases. However, as X continues to increase, Y not only stops decreasing but actually increases above its minimum value.
- **Panel E:** Exponential relationship between X and Y. In this case, Y decreases very rapidly as X first increases, but it then decreases much less rapidly as X increases further.
- **Panel F:** Little or no pattern to the relationship between X and Y. The same or similar values of Y are associated with each value of X. An X value in this data set cannot be used to predict a Y value.

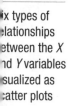

FIGURE 10.1

Six types of relationships between the X and Y variables visualized as scatter plots

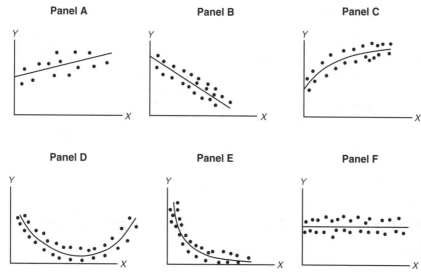

Panel A   Panel B   Panel C

Panel D   Panel E   Panel F

Simple Linear Regression

CONCEPT   The regression model that uses a straight-line, or *linear*, relationship to predict a *numerical* dependent variable Y from a single *numerical* independent variable X.

INTERPRETATION   Two values define the straight-line relationship: the Y *intercept* and the *slope*. The Y intercept is the value of the Y variable when $X = 0$. The slope is the change in the Y variable per unit change in the X variable as the chart at right shows.

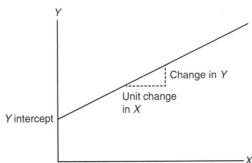

Positive straight-line relationships, which Figure 10.1 Panel A and the above chart shows, have a positive slope; negative straight-line relationships, which Figure 10.1 Panel B shows, have a negative slope.

The Simple Linear Regression Equation

CONCEPT   The equation that represents the straight-line regression model that can be used for prediction. In this equation, the symbol b_0 represents the Y intercept, and the symbol b_1 represents the slope.

INTERPRETATION   The general form of the simple linear regression equation is $Y = b_0 + b_1 X$. This means that, for a given value of X, that value is multiplied by b_1 (the slope), and the product that results is added to b_0 (the Y intercept) to calculate the predicted Y value.

The general form of the simple linear regression equation is sometimes written as $Y = a + bX$, in which the symbol a represents the Y intercept, and the symbol b represents the slope. Regardless of the way the equation is written, a simple linear regression equation is valid only for the range of X values that was used to develop the regression model, as the next section explains.

WORKED-OUT PROBLEM 1   For the simple linear regression model that the equation $Y = 100 + 1.5X$ defines for the range of X values from 10 to 50, predict the Y value that would be associated with the X value 40. The predicted Y value would be 160 ($100 + 1.5 \times 40$).

10.2 Developing a Simple Linear Regression Model

Developing a simple linear regression model requires two tasks: using a calculational method to determine the most appropriate equation and performing a residual analysis that evaluates whether a linear model is the most appropriate type of model to use.

Least-Squares Method

CONCEPT The calculational method that minimizes the sum of the squared differences between the actual values of the dependent variable Y and the predicted values of Y.

INTERPRETATION For plotted sets of X and Y values, there are many possible straight lines, each with its own values of b_0 and b_1, that might seem to fit the data. The least-squares method finds the values for the Y intercept and the slope that makes the sum of the squared differences between the actual values of the dependent variable Y and the predicted values of Y as small as possible.

Calculating the Y intercept and the slope using the least-squares method is tedious and can be subject to rounding errors if you use a simple handheld calculator. You can get more accurate results faster if you use regression software routines to perform the calculations.

WORKED-OUT PROBLEM 2 A motion picture distributor seeks to use YouTube trailer views to predict the opening weekend box office gross for movies. Such a prediction would help determine the number of theaters to use when a movie opens and would also help local theaters establish proper staffing levels. Toward this goal, an analyst collects data for 50 movies released during a one-year period from April 2019 to March 2020. The analyst collects the millions of YouTube trailer views of each movie through the Saturday before a movie opens and the opening weekend box office gross (in $millions).

Movies

Figure 10.2 presents the scatter plot for these data. The plot suggests a positive straight-line relationship between trailer views (X) and the opening weekend box office gross (Y). As the trailer views increase, opening weekend box office gross also increases.

FIGURE 10.2

Scatter plot for the motion picture distributor study (Data extracted from "Box Office Report," https://bit.ly/iQPWst.)

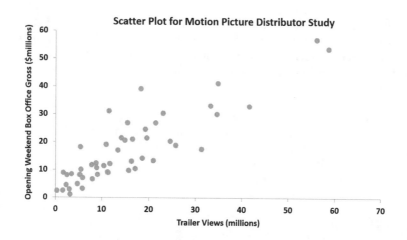

Figure 10.3 presents the spreadsheet least-squares method results for the motion picture distributor study. The results show that $b_1 = 0.8481$ and $b_0 = 3.9309$, making the equation for the best linear model for these data:

$$\text{Predicted value of opening weekend box office gross} = 3.9309 + 0.8481 \times \text{trailer views}$$

FIGURE 10.3

Simple linear regression model for the motion picture distributor study

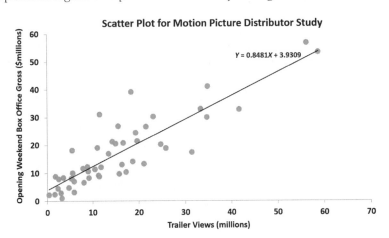

Figure 10.4 presents a revised scatter plot that now includes the plot of the simple linear regression equation determined by the Figure 10.3 results.

FIGURE 10.4

Scatter plot for the motion picture distributor study, revised to include the simple linear regression line and equation

For this regression model, the slope, b_1, is +0.8481. This means that for each increase of 1 unit in X, the value of Y is estimated to increase by 0.8481 units. In other words, for each increase of 1 million trailer views, the fitted model predicts that the opening weekend box office gross is estimated to increase by $0.8481 million.

For this regression model, the Y intercept, b_0, is 3.9309. The Y intercept represents the value of Y when X equals 0. This means that a movie with no trailer views, which is unlikely, is predicted to have first weekend sales of $3.9309 million.

Prediction Using a Simple Linear Regression Model

Once a regression model is developed, you can use the model for predicting values of a dependent variable Y from the independent variable X. However, you are restricted to the range of the values for the independent variable X. You should not extrapolate beyond the range of X values. For example, when you use the model developed in WORKED-OUT PROBLEM 2, predictions of the opening weekend box office gross should be made only for trailer views that are between 0.312 and 58.663 million.

WORKED-OUT PROBLEM 3   Using the regression model developed in WORKED-OUT PROBLEM 2, you want to predict the opening weekend box office gross for a movie that has 20 million trailer views. You predict that the opening weekend box office gross for that movie would be 20.8929 million dollars (3.9309 + 0.8481 × 20).

You use the symbols for the Y intercept, b_0, and the slope, b_1, the sample size, n, and these symbols:

- the subscripted YHat, $\hat{Y}_i$, for predicted Y values

- the subscripted italic uppercase X for the independent X values

- the subscripted italic uppercase Y for the dependent Y values

- $\overline{X}$ for the mean of the X values

- $\overline{Y}$ for the mean of the Y values

to write the equation for a simple linear regression model:

$$\hat{Y}_i = b_0 + b_1 X_i$$

You use this equation and these summations:

- $\sum_{i=1}^{n} X_i$, the sum of the X values

- $\sum_{i=1}^{n} Y_i$, the sum of the Y values

- $\sum_{i=1}^{n} X_i^2$, the sum of the squared X values

- $\sum_{i=1}^{n} X_i Y_i$, the sum of the cross-product of X and Y

to define the equation of the slope, b_1, as

$$b_1 = \frac{SSXY}{SSX}$$

in which

$$SSXY = \sum_{i=1}^{n}(X_i - \overline{X})(Y_i - \overline{Y}) = \sum_{i=1}^{n} X_i Y_i - \frac{\left(\sum_{i=1}^{n} X_i\right)\left(\sum_{i=1}^{n} Y_i\right)}{n}$$

and

$$SSX = \sum_{i=1}^{n}(X_i - \overline{X})^2 = \sum_{i=1}^{n} X_i^2 - \frac{\left(\sum_{i=1}^{n} X_i\right)^2}{n}$$

These equations, in turn, allow you to define the Y intercept as

$$b_0 = \overline{Y} - b_1 \overline{X}$$

The complexity of regression calculations invites errors when these calculations are done by handheld calculators. Therefore, all regression calculations should be done by software such as the Excel regression LINEST function that the spreadsheets for WORKED-OUT PROBLEMS 1 and 2 use.

Regression Assumptions

important point

The assumptions necessary for performing regression analysis are as follows:

- Normality of the variation around the line of regression
- Equality of variation in the Y values for all values of X
- Independence of the variation around the line of regression

Normality requires that the variation around the line of regression is normally distributed at each value of X. Like the t test and the ANOVA F test, regression analysis is fairly insensitive to departures from the normality assumption. As long as the distribution of the variation around the line of regression at each level of X is not extremely different from a normal distribution, inferences about the line of regression and the regression coefficients will not be seriously affected.

Equality of variation requires that the variation around the line of regression be constant for all values of X. This means that the variation is the same when X is a low value as when X is a high value. The equality of variation assumption is important for using the least-squares method of determining the regression coefficients. If there are serious departures from this assumption, other methods (see reference 5) can be used.

The independence of the variation around the line of regression requires that the variation around the regression line be independent for each value of X. This assumption is particularly important when data are collected over a period of time. In such situations, the variation around the line for a specific time period is often correlated with the variation of the previous time period.

Residual Analysis

Residual analysis evaluates whether a regression model that has been fitted to the data is an appropriate model. Residual analysis also helps determine whether there are violations of the regression assumptions of the regression model.

Residual

CONCEPT The difference between the observed and predicted values of the dependent variable Y for a given value of the independent variable X.

INTERPRETATION To evaluate the aptness of the fitted model, you plot the residuals on the vertical axis against the corresponding values of the independent variable on the horizontal axis. If the fitted model is appropriate for the data, there will be no apparent pattern in this plot. However, if the fitted model is not appropriate, there will be a clear relationship between values of the independent variable and the residuals.

Figure 10.5 on page 220 presents the residual plot for the motion picture distributor study simple linear regression model. In this figure, the trailer views are plotted on the horizontal X axis, and the residuals are plotted on the vertical Y axis. You see that although there is widespread scatter in the residual plot, no apparent pattern or relationship exists between the residuals and X values. The residuals appear to be evenly spread above and below 0 for the differing values of X. This observation enables you to conclude that the fitted straight-line model is appropriate for the motion picture distributor study data.

FIGURE 10.5

Residual plot for the motion picture distributor study

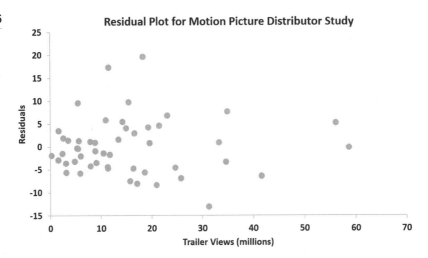

Residual Plot for Motion Picture Distributor Study

Evaluating the Assumptions

Different techniques, all involving residuals, enable you to evaluate the regression assumptions.

For equality of variation, you use the same plot as you did to evaluate the aptness of the fitted model. For the Figure 10.5 motion picture distributor study residual plot, there do not appear to be major differences in the variability of the residuals for different X values. You can conclude that for this fitted model, there is no apparent violation in the assumption of equal variation at each level of X.

For the normality of the variation around the line of regression, you plot the residuals in a histogram (see Section 2.2), boxplot (see Section 3.4), or normal probability plot (see Section 5.4). Figure 10.6 presents the histogram for the motion picture distributor study data. You see that the data appear to be approximately normally distributed, with most of the residuals concentrated in the center of the distribution.

FIGURE 10.6

Histogram of the residuals for the motion picture distributor study

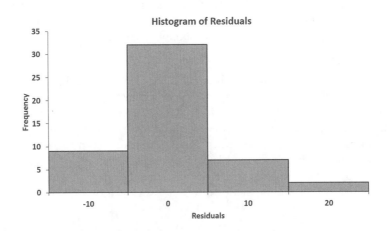

Histogram of Residuals

For the independence of the variation around the line of regression, you plot the residuals in the order in which the observed data were obtained, looking for a relationship between consecutive residuals. If you can see such a relationship, the assumption of independence is violated. Because these data were not collected over time, you do not need to evaluate this assumption.

10.3 Measures of Variation

After a regression model has been fitted to a set of data, three measures of variation determine how much of the variation in the dependent variable Y can be explained by variation in the independent variable X.

Regression Sum of Squares (SSR)

CONCEPT   The variation that is due to the relationship between X and Y.

INTERPRETATION   The regression sum of squares (SSR) is equal to the sum of the squared differences between the Y values that are predicted from the regression equation and the mean value of Y:

$$SSR = \text{Sum (Predicted } Y \text{ value} - \text{Mean } Y \text{ value)}^2$$

Error Sum of Squares (SSE)

CONCEPT   The variation that is due to factors other than the relationship between X and Y.

INTERPRETATION   The error sum of squares (SSE) is equal to the sum of the squared differences between each observed Y value and the predicted value of Y:

$$SSE = \text{Sum (Observed } Y \text{ value} - \text{Predicted } Y \text{ value)}^2$$

Total Sum of Squares (SST)

CONCEPT   The measure of variation of the Y_i values around their mean.

INTERPRETATION   The total sum of squares (SST) is equal to the sum of the squared differences between each observed Y value and the mean value of Y:

$$SST = \text{Sum (Observed } Y \text{ value} - \text{Mean } Y \text{ value)}^2$$

The total sum of squares is also equal to the sum of the regression sum of squares and the error sum of squares. For the original motion picture distributor study (WORKED-OUT PROBLEM 2), SSR is 5,995.8367, SSE is 1,877.8069, and SST is 7,873.6436.

In simple linear regression spreadsheet results, such as Figure 10.3, the SSR appears in cell C12, labeled as Regression; the SSE appears in cell C13, labeled as Residual; and the SST appears in cell C14, labeled as Total.

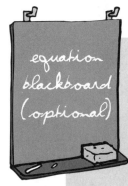

equation blackboard (optional)

interested in math?

You use symbols introduced earlier in this chapter to write the equations for the measures of variation in a regression analysis.

The equation for total sum of squares SST can be expressed in two ways:

$$SST = \sum_{i=1}^{n}(Y_i - \overline{Y})^2 \text{ equivalent to } \sum_{i=1}^{n}Y_i^2 - \frac{\left(\sum_{i=1}^{n}Y_i\right)^2}{n}$$

or as

$$SST = SSR + SSE$$

The equation for the regression sum of squares (SSR) is

$$SSR = \sum_{i=1}^{n}(\hat{Y} - \overline{Y})^2$$

which is equivalent to

$$= b_0 \sum_{i=1}^{n}Y_i + b_1 \sum_{i=1}^{n}X_iY_i - \frac{\left(\sum_{i=1}^{n}Y_i\right)^2}{n}$$

The equation for the error sum of squares (SSE) is

$$SSE = \sum_{i=1}^{n}(Y_i - \hat{Y}_i)^2$$

which is equivalent to

$$= \sum_{i=1}^{n}Y_i^2 - b_0 \sum_{i=1}^{n}Y_i - b_1 \sum_{i=1}^{n}X_iY_i$$

For the motion picture distributor study,

$$SST = \sum_{i=1}^{n}(Y_i - \bar{Y})^2 = \sum_{i=1}^{n}Y_i^2 - \frac{\left(\sum\limits_{i=1}^{n}Y_i\right)^2}{n}$$

$$= 22{,}392.13 - \frac{(852.012)^2}{50}$$

$$= 7{,}873.644$$

SSR = regression sum of squares

$$= \sum_{i=1}^{n}(\hat{Y}_i - \bar{Y})^2$$

$$= b_0\sum_{i=1}^{n}Y_i + b_1\sum_{i=1}^{n}X_iY_i - \frac{\left(\sum\limits_{i=1}^{n}Y_i\right)^2}{n}$$

$$= (3.9309)(852.012) + (0.8481)(20{,}238.47) - \frac{(852.012)^2}{50}$$

$$= 5{,}994.934$$

SSE = error sum of squares

$$= \sum_{i=1}^{n}(Y_i - \hat{Y}_i)^2$$

$$= \sum_{i=1}^{n}Y_i^2 - b_0\sum_{i=1}^{n}Y_i - b_1\sum_{i=1}^{n}X_iY_i$$

$$= 22{,}392.13 - (-3.9309)(852.012) - (0.8481)(20{,}238.47)$$

$$= 1{,}878.709$$

The SSR and SSE calculations differ slightly from the spreadsheet results due to rounding errors.

The Coefficient of Determination

CONCEPT   The ratio of the regression sum of squares to the total sum of squares, represented by the symbol r^2.

INTERPRETATION   By themselves, SSR, SSE, and SST provide little that can be directly interpreted. The ratio of the regression sum of squares (SSR) to the total sum of squares (SST) measures the proportion of variation in Y that is explained by the independent variable X in the regression model. The ratio can be expressed as follows:

$$r^2 = \frac{\text{Regression sum of squares}}{\text{Total sum of squares}} = \frac{SSR}{SST}$$

For the motion picture distributor study, $SSR = 5,995.8367$ and $SST = 7,873.6436$. Therefore,

$$r^2 = \frac{5,995.8367}{7,873.6436} = 0.7615$$

This result for r^2 means that 76.15% of the variation in the opening weekend box office gross can be explained by the variability in the trailer view. This shows a strong positive linear relationship between the two variables because the use of a regression model has reduced the variability in predicting the opening weekend box office gross by 76.15%. Only 23.85% of the sample variability in the opening weekend box office gross can be explained by factors other than what is accounted for by the linear regression model that uses only trailer views.

The Coefficient of Correlation

CONCEPT   The measure of the strength of the linear relationship between two variables, represented by the symbol r.

INTERPRETATION   The values of this coefficient vary from -1, which indicates perfect negative correlation, to $+1$, which indicates perfect positive correlation. The sign of the correlation coefficient, r, is the same as the sign of the slope in simple linear regression. If the slope is positive, r is positive. If the slope is negative, r is negative. The coefficient of correlation (r) is the square root of the coefficient of determination, r^2.

In spreadsheet results, Microsoft Excel labels the coefficient of correlation "Multiple r." For the motion picture distributor study, the coefficient of correlation, r, is $+0.8726$, the positive (because the slope is positive) square root of r^2, 0.7615. Because the coefficient is relatively close to $+1.0$, you can say that the relationship between trailer views and the opening weekend box office gross is strong. You can plausibly conclude that an increase in trailer views is associated with an increase in the opening weekend box office gross.

important point

In general, you must remember that just because two variables are strongly correlated, you cannot automatically conclude that a cause-and-effect relationship exists between those variables.

Standard Error of the Estimate

CONCEPT   The standard deviation around the fitted line of regression that measures the variability of the actual Y values from the predicted Y, represented by the symbol S_{YX}.

INTERPRETATION   Although the least-squares method results in the line that fits the data with the minimum amount of variation, unless the coefficient of determination $r^2 = 1.0$, the regression equation is not a perfect predictor.

Just as the standard deviation measures variability around the mean, the standard error of the estimate measures variability around the fitted line of regression. The second scatter plot for WORKED-OUT PROBLEM 2 shows the variability around the line of regression. In that plot, some values are above the line of regression, and other values are below the line of regression.

For the motion picture distributor study, the standard error of the estimate, which Microsoft Excel labels Standard Error in the spreadsheet results, is $6.2547 million. As Section 10.4 discusses, the standard error of the estimate can be used to determine whether a statistically significant relationship exists between the independent and dependent variables.

equation blackboard (optional)

interested in math?

You use symbols introduced earlier in this chapter to write the equation for the standard error of the estimate:

$$S_{YX} = \sqrt{\frac{SSE}{n-2}} = \sqrt{\frac{\sum_{i=1}^{n}(Y_i - \hat{Y}_i)^2}{n-2}}$$

For the motion picture distributor study, with SSE equal to 1,877.8069,

$$S_{YX} = \sqrt{\frac{1,877.8069}{50-2}}$$

$$S_{YX} = 6.2547$$

10.4 Inferences About the Slope

Once you have verified that the assumptions of the least-squares regression model have not been seriously violated and that a straight-line model is appropriate, you can use a simple linear regression model for statistical inference. The regression model that is based on sample data can now be used to make inferences about the population. The *t* test for the slope and developing a confidence interval estimate of the slope are two inferences commonly encountered.

t Test for the Slope

You can determine the existence of a significant relationship between the X and Y variables by testing whether β_1 (the population slope) is equal to 0. If this hypothesis is rejected, you conclude that evidence of a linear relationship exists. The null and alternative hypotheses are as follows:

$H_0: \beta_1 = 0$ (No linear relationship exists.)

$H_1: \beta_1 \neq 0$ (A linear relationship exists.)

The test statistic follows the *t* distribution with the degrees of freedom equal to the sample size minus 2. The test statistic is equal to the sample slope divided by the standard error of the slope:

$$t = \frac{\text{sample slope}}{\text{standard error of the slope}}$$

For the motion picture distributor study, the critical value of *t* at the $\alpha = 0.05$ level of significance is 2.0106, the value of *t* is 12.38, and the *p*-value is 0.0000. (In the spreadsheet results, Microsoft Excel labels the *t* statistic as t Stat.) Using the *p*-value approach, you reject H_0 because the *p*-value 0.0000 is less than $\alpha = 0.05$. Using the critical value approach, you reject H_0 because $t = 12.38 > 2.0106$. You can conclude that a significant linear relationship exists between the trailer views and the opening weekend box office gross.

Confidence Interval Estimate of the Slope (β_1)

You can also test the existence of a linear relationship between variables by constructing a confidence interval estimate of β_1 and determining whether the hypothesized value ($\beta_1 = 0$) is included in the interval.

You construct the confidence interval estimate of the slope β_1 by multiplying the *t* statistic by the standard error of the slope and then adding and subtracting this product to the sample slope.

For the motion picture distributor study, the spreadsheet results include the calculated lower and upper limits of the confidence interval estimate for the slope of the trailer views and opening weekend box office gross. With 95% confidence, the lower limit is 0.7104, and the upper limit is 0.9859.

Because these values are above 0, you conclude that a significant linear relationship exists between the trailer views and opening weekend box office gross. The confidence interval indicates that for each increase of 1 million trailer views, the opening weekend box office gross is estimated to increase by at least $0.7104 million but less than $0.9859 million. Had the interval included 0, you would have concluded that no relationship exists between the variables.

equation blackboard (optional)

interested in math?

You assemble symbols introduced earlier and the symbol for the standard error of the slope, S_{b_1}, to form the equation for the t statistic used in testing a hypothesis for a population slope β_1.

You begin by forming the equations for the standard error of the slope, S_{b_1} as

$$S_{b_1} = \frac{S_{YX}}{\sqrt{SSX}}$$

Then, you use the standard error of the slope, S_{b_1} to define the test statistic:

$$t = \frac{b_1 - \beta_1}{S_{b_1}}$$

The test statistic t follows a t distribution with $n - 2$ degrees of freedom.

For the motion picture distributor study, to test whether a significant relationship exists between the trailer views and the opening weekend box office gross at the level of significance $\alpha = 0.05$,

$$b_1 = +0.8481 \qquad S_{b_1} = 0.0685$$

$$t = \frac{b_1 - \beta_1}{S_{b_1}}$$

$$= \frac{0.8481 - 0}{0.0685} = 12.375$$

spreadsheet solution

Simple Linear Regression

Chapter 10 Simple Linear Regression contains the spreadsheet that WORKED-OUT PROBLEM 2 uses to create a simple linear regression model for the motion picture distributor study data.

Best Practices

Use the **LINEST(***cell range of Y variable, cell range of X variable***, True, True)** function to calculate the b_0 and b_1 coefficients and standard errors, r^2, the standard error of the estimate, the F test statistic, the residual degrees of freedom, and *SSR* and *SSE*.

Use the **T.INV.2T(***1 − confidence level, residual degrees of freedom***)** function to calculate the critical value for the t test.

To calculate Significance F, the p-value for the F test, use the **F.DIST.RT(***F critical value, regression degrees of freedom, residual degrees of freedom***)** function.

To calculate the p-values in the second part of the spreadsheet ANOVA table, use the **T.DIST.2T(***absolute value of the t test statistic, residual degrees of freedom***)** function.

How-Tos

Experiment with this spreadsheet by changing the confidence level of significance in cell L8.

Examine the **Chapter 10 Simple Linear Regression** RESIDUALS spreadsheet, which does not appear in this chapter, for a model for calculating residuals.

Advanced Technique ATT6 in Appendix E describes using the Analysis ToolPak as a second way to perform regression analysis.

Advanced Technique ADV6 in Appendix E explains more about how to use the **LINEST** function to calculate regression results.

Advanced Technique ADV7 in Appendix E explains how to modify **Chapter 10 Simple Linear Regression** for use with other data sets.

equation blackboard (optional)

You assemble symbols introduced earlier to form the equation for the confidence interval estimate of the slope β_1:

$$b_1 \pm t_{n-2} S_{b_1}$$

For the movie company example,

$$b_1 = +0.8481 \qquad n = 50 \qquad S_{b_1} = 0.0685$$

Using 95% confidence, with degrees of freedom = $50 - 2 = 48$,

$$
\begin{aligned}
b_1 \pm t_{n-2} S_{b_1} \\
&= +0.8481 \pm (2.0106)(0.0685) \\
&= +0.8481 \pm 0.1378 \\
& \quad +0.7103 \le \beta_1 \le +0.9859
\end{aligned}
$$

interested in math?

10.5 Common Mistakes When Using Regression Analysis

Some of the common mistakes that people make when using regression analysis are as follows:

important point

- Lacking awareness of the assumptions of least-squares regression
- Not knowing how to evaluate the assumptions of least-squares regression
- Not knowing what the alternatives to least-squares regression are if a particular assumption is violated
- Using a regression model without knowledge of the subject of the analysis
- Predicting Y outside the relevant range of X

Most software regression analysis routines do not check for these mistakes. You must always use regression analysis wisely and always check that others who provide you with regression results have avoided these mistakes as well.

The following four sets of data illustrate some of the mistakes that you can make in regression analysis.

Anscombe

Data Set A		Data Set B		Data Set C		Data Set D	
X_i	Y_i	X_i	Y_i	X_i	Y_i	X_i	Y_i
10	8.04	10	9.14	10	7.46	8	6.58
14	9.96	14	8.10	14	8.84	8	5.76
5	5.68	5	4.74	5	5.73	8	7.71
8	6.95	8	8.14	8	6.77	8	8.84
9	8.81	9	8.77	9	7.11	8	8.47
12	10.84	12	9.13	12	8.15	8	7.04
4	4.26	4	3.10	4	5.39	8	5.25
7	4.82	7	7.26	7	6.42	19	12.50
11	8.33	11	9.26	11	7.81	8	5.56
13	7.58	13	8.74	13	12.74	8	7.91
6	7.24	6	6.13	6	6.08	8	6.89

Source: Extracted from F. J. Anscombe, "Graphs in Statistical Analysis," *American Statistician*, Vol. 27 (1973), pp. 17–21.

Anscombe (reference 1) showed that for the four data sets, the regression results are identical:

$$\text{predicted value of } Y = 3.0 + 0.5X_i$$
$$\text{standard error of the estimate} = 1.237$$
$$r^2 = 0.667$$
$$SSR = \text{regression sum of squares} = 27.51$$
$$SSE = \text{error sum of squares} = 13.76$$
$$SST = \text{total sum of squares} = 41.27$$

However, the four data sets are actually quite different, as the Figure 10.7 scatter plots and residual plots for the four data sets on the next page illustrate.

Examining the Figure 10.7 plots reveals how different these four data sets are:

- **Data set A:** The scatter plot seems to follow an approximate straight line, and the residual plot does not show any obvious patterns or outlying residuals.

(*continues on page 232*)

FIGURE 10.7

Scatter plots and residual plots for the Anscombe data sets

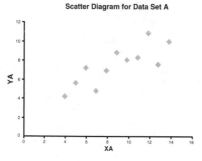

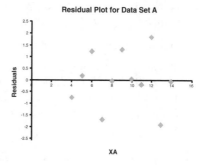

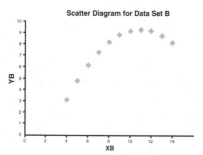

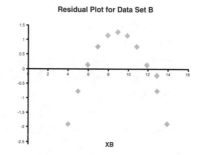

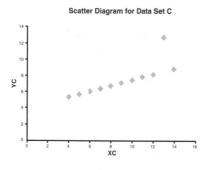

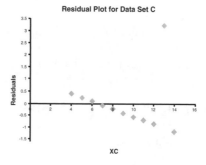

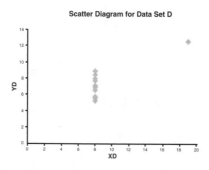

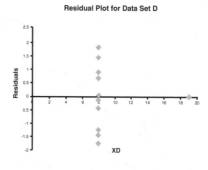

- **Data set B:** The scatter plot suggests that a curvilinear regression model should be considered, and the residual plot reinforces that suggestion.
- **Data set C:** Both the scatter and residual plots depict an extreme value.
- **Data set D:** The scatter plot reveals a single extreme value (at $X = 19$, $Y = 12.50$) that the residual plot shows greatly influences the fitted model. The regression model fit for these data should be evaluated cautiously because the regression coefficients for the model are heavily dependent on this single value.

To avoid the common mistakes of regression analysis, follow these steps:

1. Always start with a scatter plot to observe the possible relationship between X and Y.
2. Check the assumptions of regression after the regression model has been fitted, before using the results of the model.
3. Plot the residuals versus the independent variable to verify that the model fitted is an appropriate model and to see if the equal variation assumption was violated.
4. Use a histogram, boxplot, or normal probability plot of the residuals to visually evaluate whether the normality assumption has been seriously violated.
5. If the evaluation of the residuals indicates violations in the assumptions, use alternative methods to least-squares regression or alternative least-squares models (see reference 5), as appropriate.

If the evaluation of the residuals does not indicate violations in the assumptions, then you can undertake the inferential aspects of the regression analysis. You can perform a test for the significance of the slope, and you can construct a confidence interval estimate of the slope.

Important Equations

Regression Equation:

$$\hat{Y}_i = b_0 + b_1 X_i$$

Slope:

$$b_1 = \frac{SSXY}{SSX}$$

Sum of Squares:

$$SSXY = \sum_{i=1}^{n}(X_i - \bar{X})(Y_i - \bar{Y}) = \sum_{i=1}^{n}X_iY_i - \frac{\left(\sum_{i=1}^{n}X_i\right)\left(\sum_{i=1}^{n}Y_i\right)}{n}$$

and

$$SSX = \sum_{i=1}^{n}(X_i - \bar{X})^2 = \sum_{i=1}^{n}X_i^2 - \frac{\left(\sum_{i=1}^{n}X_i\right)^2}{n}$$

Y Intercept:

$$b_0 = \bar{Y} - b_1 \bar{X}$$

Total Sum of Squares:

$$SST = \sum_{i=1}^{n}(Y_i - \bar{Y})^2 \text{ equivalent to} \sum_{i=1}^{n}Y_i^2 - \frac{\left(\sum_{i=1}^{n}Y_i\right)^2}{n}$$

$$SST = SSR + SSE$$

Regression Sum of Squares:

SSR = Explained variation or regression sum of squares

$$= \sum_{i=1}^{n}(\hat{Y}_1 - \hat{Y})^2$$

which is equivalent to

$$b_0\sum_{i=1}^{n}Y_i + b_1\sum_{i=1}^{n}X_iY_i - \frac{\left(\sum_{i=1}^{n}Y_i\right)^2}{n}$$

Error Sum of Squares:

SSE = Unexplained variation or error sum of squares

$$= \sum_{i=1}^{n}(Y_i - \hat{Y}_i)^2$$

which is equivalent to

$$= \sum_{i=1}^{n}Y_i^2 - b_0\sum_{i=1}^{n}Y_i - b_1\sum_{i=1}^{n}X_iY_i$$

Coefficient of Determination:

$$r^2 = \frac{\text{regression sum of squares}}{\text{total sum of squares}} = \frac{SSR}{SST}$$

Coefficient of Correlation:

$$r = \sqrt{r^2}$$

If b_1 is positive, r is positive. If b_1 is negative, r is negative.

Standard Error of the Estimate:

$$S_{YX} = \sqrt{\frac{SSE}{n-2}} = \sqrt{\frac{\sum_{i=1}^{n}(Y_i - \hat{Y}_i)^2}{n-2}}$$

t Test for the Slope:

$$t = \frac{b_1 - \beta_1}{S_{b_1}}$$

One-Minute Summary

Simple linear regression

- Least-squares method
- Measures of variation
- Residual analysis
- t test for the significance of the slope
- Confidence interval estimate of the slope

Test Yourself
Short Answers

1. The Y intercept (b_0) represents the:
   a. predicted value of Y when $X = 0$
   b. change in Y per unit change in X
   c. predicted value of Y
   d. variation around the regression line

2. The slope (b_1) represents the:

   a. predicted value of Y when $X = 0$
   b. change in Y per unit change in X
   c. predicted value of Y
   d. variation around the regression line

3. The standard error of the estimate is a measure of:

   a. total variation of the Y variable
   b. the variation around the regression line
   c. explained variation
   d. the variation of the X variable

4. The coefficient of determination (r^2) tells you:

   a. that the coefficient of correlation (r) is larger than 1
   b. whether the slope has any significance
   c. whether the regression sum of squares is greater than the total sum of squares
   d. the proportion of total variation that is explained

5. In performing a regression analysis involving two numerical variables, you assume that:

   a. the variances of X and Y are equal
   b. the variation around the line of regression is the same for each X value
   c. X and Y are independent
   d. All of the above

6. Which of the following assumptions concerning the distribution of the variation around the line of regression (the residuals) is correct?

   a. The distribution is normal.
   b. All of the variations are positive.
   c. The variation increases as X increases.
   d. Each residual is dependent on the previous residual.

7. The residuals represent:

   a. the difference between the actual Y values and the mean of Y
   b. the difference between the actual Y values and the predicted Y values
   c. the square root of the slope
   d. the predicted value of Y when $X = 0$

8. If the coefficient of determination (r^2) = 1.00, then:

   a. the Y intercept must equal 0
   b. the regression sum of squares (SSR) equals the error sum of squares (SSE)
   c. the error sum of squares (SSE) equals 0
   d. the regression sum of squares (SSR) equals 0

9. If the coefficient of correlation $(r) = -1.00$, then:

   a. All the data points must fall exactly on a straight line with a slope that equals 1.00.

   b. All the data points must fall exactly on a straight line with a negative slope.

   c. All the data points must fall exactly on a straight line with a positive slope.

   d. All the data points must fall exactly on a horizontal straight line with a zero slope.

10. Assuming a straight-line (linear) relationship between X and Y, if the coefficient of correlation (r) equals -0.30:

    a. there is no correlation

    b. the slope is negative

    c. variable X is larger than variable Y

    d. the variance of X is negative

11. The strength of the linear relationship between two numerical variables is measured by the:

    a. predicted value of Y

    b. coefficient of determination

    c. total sum of squares

    d. Y intercept

12. In a simple linear regression model, the coefficient of correlation and the slope:

    a. may have opposite signs

    b. must have the same sign

    c. must have opposite signs

    d. are equal

Answer True or False:

13. The regression sum of squares (SSR) can never be greater than the total sum of squares (SST).

14. The coefficient of determination represents the ratio of SSR to SST.

15. Regression analysis is used for prediction, while correlation analysis is used to measure the strength of the association between two numerical variables.

16. The value of r is always positive.

17. When the coefficient of correlation $r = -1$, a perfect relationship exists between X and Y.

18. If no apparent pattern exists in the residual plot, the regression model fit is appropriate for the data.

19. If the range of the X variable is between 100 and 300, you should not make a prediction for $X = 400$.

20. If the p-value for a t test for the slope is 0.021, the results are significant at the 0.01 level of significance.

Fill in the Blank:

21. The residual represents the difference between the observed value of Y and the _____ value of Y.

22. The change in Y per unit change in X is called the _____.

23. The ratio of the regression sum of squares (SSR) to the total sum of squares (SST) is called the _____.

24. In simple linear regression, if the slope is positive, then the coefficient of correlation must also be _____.

25. One of the assumptions of regression is that the residuals around the line of regression follow the _____ distribution.

Answers to Test Yourself Short Answers

1. a
2. b
3. b
4. d
5. b
6. a
7. b
8. c
9. b
10. b
11. b
12. b
13. True

14. True
15. True
16. False
17. True
18. True
19. True
20. False
21. predicted
22. slope
23. coefficient of determination
24. positive
25. normal

Problems

1. The asking price (in thousands of dollars) and property size (in square feet) were collected for a sample of 62 single-family homes located in Glen Cove, New York. Develop a simple linear regression model to predict asking price based on the property size.

Glen Cove

a. Assuming a linear relationship, use the least-squares method to compute the regression coefficients b_0 and b_1. State the regression equation for predicting the asking price based on the property size.

b. Interpret the meaning of the Y intercept b_0 and the slope b_1 in this problem.

c. Explain why the regression coefficient, b_0, has no practical meaning in the context of this problem.

d. Predict the asking price for a house that has a property size of 14,000 square feet.

e. Compute the coefficient of determination, r^2, and interpret its meaning.

f. Perform a residual analysis on the results and determine the adequacy of the model.

g. Determine whether a significant relationship exists between asking price and the property size at the 0.05 level of significance.

h. Construct a 95% confidence interval estimate of the population slope between the asking price and the property size.

2. Measuring the height of a California redwood tree is a very difficult undertaking because these trees grow to heights of more than 300 feet. People familiar with these trees understand that the height of a California redwood tree is related to other characteristics of the tree, including the diameter of the tree at the breast height of a person. The following data represent the height (in feet) and diameter at breast height of a person for a sample of 21 California redwood trees.

Redwood

Height	Diameter at Breast Height	Height	Diameter at Breast Height
122.0	20	164.0	40
193.5	36	203.3	52
166.5	18	174.0	30
82.0	10	159.0	22
133.5	21	205.0	42
156.0	29	223.5	45
172.5	51	195.0	54
81.0	11	232.5	39
148.0	26	190.5	36
113.0	12	100.0	8
84.0	13		

a.   Assuming a linear relationship, use the least-squares method to compute the regression coefficients b_0 and b_1. State the regression equation that predicts the height of a tree based on the tree's diameter at breast height of a person.

b.   Interpret the meaning of the slope in this equation.

c.   Predict the height of a tree that has a diameter at breast height of 25 inches.

d.   Interpret the meaning of the coefficient of determination in this problem.

e.   Perform a residual analysis on the results and determine the adequacy of the model.

f.   Determine whether a significant relationship exists between the height of redwood trees and the diameter at breast height at the 0.05 level of significance.

g.   Construct a 95% confidence interval estimate of the population slope between the height of the redwood trees and diameter at breast height.

3. A baseball analyst would like to study various team statistics from a recent baseball season to determine which variables might be useful in predicting the winning percentage achieved by teams during the season. He has decided to begin by using a team's earned run average (ERA), a measure of pitching performance, to predict the winning percentage.

(Hint: First, determine which are the independent and dependent variables.)

Baseball

a.   Assuming a linear relationship, use the least-squares method to compute the regression coefficients b_0 and b_1. State the regression equation for predicting the winning percentage, based on the ERA.

b.   Interpret the meaning of the Y intercept, b_0, and the slope, b_1, in this problem.

c.   Predict the winning percentage for a team with an ERA of 3.75.

d.   Compute the coefficient of determination, r^2, and interpret its meaning.

e.   Perform a residual analysis on your results and determine the adequacy of the fit of the model.

f.   At the 0.05 level of significance, does evidence exist of a linear relationship between the winning percentage and the ERA?

g.   Construct a 95% confidence interval estimate of the slope.

h.   What other independent variables might you include in the model?

4. *Consumer Reports* publishes information about automobiles. From that source, data on the miles per gallon (MPG), weight (in pounds), and acceleration from 0 to 30 miles per hour (in seconds) for a sample of 26 2021 sedans were collected. Using that sample, develop a regression model to predict the miles per gallon based on the weight of the automobile.

MPG

a.   Assuming a linear relationship, use the least-squares method to compute the regression coefficients b_0 and b_1. State the regression equation for predicting miles per gallon based on the weight of a sedan.

b.   Interpret the meaning of the Y intercept, b_0, and the slope, b_1, in this problem.

c.   Predict the miles per gallon for an automobile that weighs 3,700 pounds.

d.   Compute the coefficient of determination, r^2, and interpret its meaning.

e.   Perform a residual analysis on your results and determine the adequacy of the fit of the model.

f.   At the 0.05 level of significance, does evidence exist of a linear relationship between the miles per gallon and the weight of a sedan?

g.   Construct a 95% confidence interval estimate of the population slope between the miles per gallon and the weight of a sedan.

h.   How useful do you think the weight is as a predictor of miles per gallon? Explain.

Answers to Problems

1. a. $b_0 = 514.2037$, $b_1 = 0.0225$; Predicted asking price $= 514.2037 + 0.0225$ sq. ft.

b.   Each increase by 1 square foot in property size is estimated to increase value by $0.0225.

c.   The interpretation of b_0 has no practical meaning here because it would represent the estimated asking price of a house that has no property size.

d.   Predicted asking price $= 514.2037 + 0.0225 (14,000) = \$828.7078$ thousands.

e. $r^2 = 0.5343$. 53.43% of the variation in the asking price of a house can be explained by variation in property size.

f.   There is no particular pattern in the residual plot, and the model appears to be adequate.

g. $t = 8.297$; p-value $0.0000 < 0.05$ (or $t = 5.365 > 2.0003$). Reject H_0 at the 5% level of significance. There is evidence of a significant linear relationship between asking price and property size.

h. $0.0170 < \beta_1 < 0.0279$.

2. a. $b_0 = 78.7963$, $b_1 = 2.6732$; predicted height $= 78.7963 + 2.6732$ diameter of the tree at breast height of a person (in inches).

   b.   For each additional inch in the diameter of the tree at breast height of a person, the height of the tree is estimated to increase by 2.6732 feet.

   c.   Predicted height $= 78.7963 + 2.6732\,(25) = 145.6267$ feet.

   d. $r^2 = 0.7288$. 72.88% of the total variation in the height of the tree can be explained by the variation of the diameter of the tree at breast height of a person.

   e.   There is no particular pattern in the residual plot, and the model appears to be adequate.

   f. $t = 7.1455$; p-value = virtually $0 < 0.05$ ($t = 7.1455 > 2.093$). Reject H_0. There is evidence of a significant linear relationship between the height of the tree and the diameter of the tree at breast height of a person.

   g. $1.8902 < \beta_1 < 3.4562$.

3. a. $b_0 = 0.9839$, $b_1 = -0.1088$; predicted winning percentage $= 0.9839 - 0.1088$ ERA.

   b.   For each additional earned run allowed, the winning percentage is estimated to decrease by 0.1088.

   c.   Predicted winning percentage $= 0.9839 - 0.1088\,(3.75) = 0.5758$.

   d. $r^2 = 0.6529$. 65.29% of the total variation in the winning percentage can be explained by the variation of the ERA.

   e.   There is no particular pattern in the residual plot, and the model appears to be adequate.

   f. $t = -7.2567$; p-value = virtually $0 < 0.05$ ($t = -7.2567 < -2.0484$). Reject H_0. There is evidence of a significant linear relationship between the winning percentage and the ERA.

   g. $-0.1395 < \beta_1 < -0.0781$.

   h.   Among the independent variables, you could consider including in the model runs scored per game, hits allowed, saves, walks allowed, and errors.

4. a. $b_0 = 50.3332$, $b_1 = -0.0064$; predicted MPG $= 50.3332 - 0.0064$ weight in pounds.

   b.   For each additional increase of 1 pound, the MPG is estimated to decrease by 0.0064. Because no car will have a weight of 0, it is inappropriate to interpret the Y intercept.

   c.   Predicted MPG $= 50.3332 - 0.0064\,(3,770) = 26.1018$.

   d. $r^2 = 0.8870$. This means that 88.70% of the variation in the MPG can be explained by the variation in the weight.

e.   There is no obvious pattern in the residuals, so the assumptions of regression are met. The model appears to be adequate.

f. $t = -13.7265$; p-value = virtually 0, which is less than 0.05 (or $t = -13.7265 < -2.0639$); reject H_0. There is evidence of a linear relationship between MPG and weight.

g. $-0.0074 < \beta_1 < -0.0055$.

h.   The linear regression model appears to have provided an adequate fit and shows a significant linear relationship between MPG and weight. Because 88.70% of the variation in the MPG can be explained by the variation in weight, weight would be useful in predicting the MPG.

References

1. Anscombe, F. J. "Graphs in Statistical Analysis." *American Statistician*, 27, 1973, pp. 17–21.

2. Berenson, M. L., D. M. Levine, K. A. Szabat, and D. Stephan. *Basic Business Statistics: Concepts and Applications*, 15th edition. Hoboken, NJ: Pearson Education, 2023.

3. Hosmer, D. W., and S. Lemeshow. *Applied Logistic Regression*, 3rd edition. New York: Wiley, 2013.

4. Johnson, S. "The Trouble with QSAR (or How I Learned to Stop Worrying and Embrace Fallacy)." *Journal of Chemical Information and Modeling*, 48, 2008, pp. 25–26.

5. Kutner, M. H., C. Nachtsheim, J. Neter, and W. Li. *Applied Linear Statistical Models*, 5th edition. New York: McGraw-Hill-Irwin, 2005.

6. Levine, D. M., D. Stephan, and K. A. Szabat. *Statistics for Managers Using Microsoft Excel*, 9th edition. Hoboken, NJ: Pearson Education, 2021.

7. Vidakovic, B. *Statistics for Bioengineering Sciences: With MATLAB and WinBUGS Support*. New York: Springer Science+Business Media, 2011.

CHAPTER

Multiple Regression

Chapter 10 discusses the simple linear regression model that uses one numerical independent variable. This chapter introduces you to multiple regression models that use two or more independent variables to predict the value of a dependent variable.

11.1 The Multiple Regression Model

CONCEPT The statistical method that extends the simple linear regression model linear relationship between each independent variable and the dependent variable.

EXAMPLES A regression model that predicts store sales using size of store and the mean disposable income in the area near the store, a regression model that predicts opening weekend box office gross of a movie using number of trailer views and promotional expense for a movie.

INTERPRETATION Often times you can make better predictions if you use more than one independent variable to predict a dependent variable. Because multiple regression extends simple linear regression, learning simple linear regression first (the focus of Chapter 10) helps in understanding this more advanced form of regression.

Multiple regression is one of the most commonly used data analysis methods, and this inferential method also serves as a building block for some advanced analytics methods. The independence of the independent variables used is most important because an independent variable that can predict another *independent* variable can complicate the understanding of a multiple regression model.

In a multiple regression model, each independent variable is represented by a subscripted X, such as X_1 or X_2. For a model with the two independent variables, X_1 and X_2, the general form of the multiple regression equation is

$$Predicted\ Y = b_0 + b_1X_1 + b_2X_2$$

WORKED-OUT PROBLEM 1   A real estate broker wants to develop a multiple regression model to predict the asking price of a home based on the property size and the age of a home.

Glen Cove

From the website Realtor.com, you extract a sample of 62 homes in Glen Cove, New York, and perform a multiple regression analysis using the property size in square feet (X_1) and the age of the house in years (X_2) as the independent variables to predict the dependent variable asking price (in $thousands). Figure 11.1 presents the spreadsheet multiple regression results for the sample of 62 homes.

FIGURE 11.1

Multiple regression model for the sample of 62 homes

	A	B	C	D	E	F	G	H	I	J	K	L	M	N
1	Multiple Regression Model to Predict Asking Price											Calculations		
2											b2, b1, b0 intercepts	-5.7672	0.0208	930.1169
3	Regression Statistics										b2, b1, b0 Standard Error	1.0025	0.0022	84.5721
4	Multiple R	0.8377									R Square, Standard Error	0.7017	227.2310	#N/A
5	R Square	0.7017									F, Residual df	69.3806	59	#N/A
6	Adjusted R Square	0.6915									Regression SS, Residual SS	7164791.119	3046402.817	#N/A
7	Standard Error	227.2310												
8	Observations	62												
9											Confidence level		95%	
10	ANOVA										t Critical Value		2.0010	
11		df	SS	MS	F	Significance F					Half Width b0		169.2283	
12	Regression	2	7164791.1190	3582395.5595	69.3806	0.0000					Half Width b1		0.0044	
13	Residual	59	3046402.8185	51633.9460							Half Width b2		2.0059	
14	Total	61	10211193.9355											
15														
16		Coefficients	Standard Error	t Stat	P - value	Lower 95%	Upper 95%	Lower 95%	Upper 95%					
17	Intercept	930.1169	84.5721	10.9979	0.0000	760.8886	1099.3453	760.8886	1099.3453					
18	Property Size	0.0208	0.0022	9.4571	0.0000	0.0164	0.0252	0.0164	0.0252					
19	Age	-5.7672	1.0025	-5.7529	0.0000	-7.7731	-3.7612	-7.7731	-3.7612					

Net Regression Coefficients

CONCEPT   The coefficients that measure the change in Y per unit change of a particular X, holding constant the effect of the other X variables. Net regression coefficients are also known as **partial regression coefficients**.

important point

INTERPRETATION   Multiple regression models have multiple net regression coefficients. In a multiple regression model with two independent variables, there are two net regression coefficients, in addition to the Y intercept:

Coefficient	Definition
b_0	The Y intercept (the same as in simple linear regression)
b_1	The slope of variable X_1, the change in Y per unit change in X_1, taking into account the effect of X_2
b_2	The slope of variable X_2, the change in Y per unit change in X_2, taking into account the effect of X_1

For the real estate study, b_1, the slope of Y with X_1, is 0.0208, and b_2, the slope of Y with X_2, is −5.7672. The slope b_1 means that for each increase of 1 square foot in property size (X_1), the asking price (Y) is estimated to increase by $0.0208 thousand ($20.80), holding constant the effect of X_2. The slope b_2 means that for each increase of 1 year of age (X_2), the asking price (Y) is estimated to decrease by $5.767.20 thousand ($5,767.20), holding constant the effect of X_1.

Another way to interpret the net effect is to think of two homes that are the same age. If the first home consists of 1,000 square feet more than the other home, the *net effect* of this difference is that the first home is predicted to have an asking price that is $20.80 more than the other house.

Likewise, consider two homes that are the same size in square footage. If the first home is one year older than the other home, the *net effect* of this difference is that the first home is predicted to have an asking price that is $5,767.20 less than the other house.

Predicting the Dependent Variable

As with a simple regression model, you can use a verified multiple regression model to predict values of a dependent variable, given values of the independent variables that fall inside the ranges of values that were used to create the multiple regression model. When using a multiple regression model, remember that, like a simple linear regression model, the Y intercept, b_0, may have no practical interpretation and that the model is valid only within the ranges of the independent variables.

WORKED-OUT PROBLEM 2   Using the real estate multiple regression model that WORKED-OUT PROBLEM 1 develops, you want to predict the asking price for a home that has a property size of 10,004 square feet and that is 65 years old.

For the real estate study, the Y intercept, b_0, is 930.1169. Therefore, the multiple regression equation for this study is

$$\text{Predicted value of asking price} =$$
$$930.1169 + 0.0208 \times \text{square feet} + -5.7672 \times \text{age}$$

This model is valid only for properties that are 1,422 to 79,279 square feet in size and that are 13 to 133 years old. Because 10,004 square feet and 65 years old are inside these ranges, you can use the equation to predict that the asking price will be $763,729.50 [930.1169 + (0.208 × 10,004) + (−5.7672 × 65)].

11.2 Coefficient of Multiple Determination

CONCEPT   The statistic that represents the proportion of the variation in Y that is explained by the set of independent variables included in the multiple regression model.

INTERPRETATION   The coefficient of multiple determination is analogous to the coefficient of determination (r^2) that measures the variation in Y that is explained by the independent variable X in the simple linear regression model.

WORKED-OUT PROBLEM 3   You need to calculate this coefficient for the real estate study. In the ANOVA summary table of the spreadsheet results, SSR is 7,164,791.1190, and SST is 10,211,193.9355. Therefore,

$$r^2 = \frac{\text{regression sum of squares}}{\text{total sum of squares}} = \frac{SSR}{SST} = \frac{7,164,791.1190}{10,211,193.9355} = 0.7017$$

The coefficient of multiple determination $r^2 = 0.7017$ indicates that 70.17% of the variation in asking price is explained by the variation in the property size and the age of the home.

11.3 The Overall *F* Test

CONCEPT   The test for the significance of the overall multiple regression model.

INTERPRETATION   You use this test to determine whether a significant relationship exists between the dependent variable and the entire set of independent variables. Because there is more than one independent variable, you have the following null and alternative hypotheses:

H_0: No linear relationship exists between the dependent variable and the independent variables.

H_1: A linear relationship exists between the dependent variable and at least one of the independent variables.

The ANOVA summary table for the overall F test is as follows (n = sample size and k = number of independent variables):

Source	Degrees of Freedom	Sum of Squares	Mean Square (Variance)	F
Regression	k	SSR	$MSR = \dfrac{SSR}{k}$	$F = \dfrac{MSR}{MSE}$
Error	$n - k - 1$	SSE	$MSE = \dfrac{SSE}{n - k - 1}$	
Total	$n - 1$	SST		

WORKED-OUT PROBLEM 4   You are asked to perform the overall F test for the real estate study. Using the ANOVA table in spreadsheet results for WORKED-OUT PROBLEM 1, the F statistic is 69.3806 and p-value = 0.000. Because the p-value, 0.000, is less than 0.05, you reject H_0 and conclude that at least one of the independent variables, property size or age, is related to the asking price.

11.4 Residual Analysis for the Multiple Regression Model

Residual analysis for multiple regression models is similar to residual analysis for simple linear regression. For example, for the multiple regression model with two independent variables, you analyze up to four residual plots:

- Residuals versus the predicted value of Y
- Residuals versus the first independent variable X_1
- Residuals versus the second independent variable X_2
- Residuals versus time (if the data has been collected in time order)

If the residuals versus the predicted value of Y show a pattern for different predicted values of Y, there is evidence of a possible curvilinear effect in at least one independent variable. This indicates a possible violation to the assumption of equal variance and/or the need to transform the Y variable.

Patterns in the plot of the residuals versus an independent variable, the second and third types of plots, may indicate the existence of a curvilinear effect. This indicates the need to add a curvilinear independent variable to the multiple regression model, a technique that references 1 and 2 discuss.

Patterns in the residuals in the fourth type of plot can help determine whether the independence assumption has been violated when the data are collected in time order. This plot is not done when the data being analyzed were not collected in time order, as is the case for the real estate study data.

WORKED-OUT PROBLEM 5   You are asked to perform a residual analysis for the multiple regression model for the real estate study. You create three types of residual plots for the model (see Figure 11.2).

In these plots, you see very little or no pattern in the relationship between the residuals and the predicted asking price (Y), the property size (X_1), or the age of the home (X_2). You conclude that the multiple regression model is appropriate for predicting asking price.

FIGURE 11.2

Residual plots for the real estate study

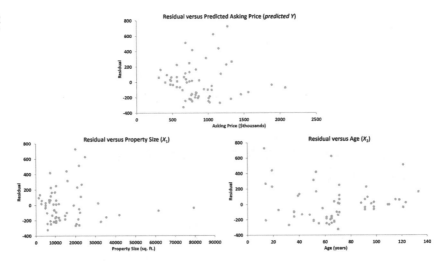

11.5 Inferences Concerning the Population Regression Coefficients

The t test for the slope and the confidence interval estimate of the slope that Chapter 10 discusses for simple linear regression models can be extended to making inferences using a verified multiple regression model.

t Test for the Slope

For each independent variable in a multiple regression model, you use a t test for the population slope. For a multiple regression model with two independent variables, the null hypothesis for each independent variable is that no linear relationship exists between the dependent variable and the independent variable, holding constant the effect of the other independent variables. The alternative hypothesis is that a linear relationship exists between the dependent variable and the independent variable, holding constant the effect of the other independent variables.

This t test serves a second purpose as a test for the significance of adding a particular independent variable into a multiple regression model, given the presence of another variable already in the model.

WORKED-OUT PROBLEM 6   You want to perform a t test for the slope for the real estate study multiple regression model. From the Figure 11.1 spreadsheet results, for the property size independent variable, the t statistic is 9.4571 and the p-value is 0.000. Because the p-value, 0.000, is less than 0.05, you reject the null hypothesis and conclude that a linear relationship exists between the asking price and the property size (X_1).

For the age independent variable, the t statistic is -5.7529 and the p-value is 0.000. Because p-value, 0.000, is less than 0.05, you reject the null hypothesis and conclude that a linear relationship exists between the asking price and the age of a home (X_2). You conclude that because each of the two independent variables is significant, both should be included in the regression model.

Confidence Interval Estimate of the Slope

You can construct a confidence interval estimate of the slope for a multiple regression model. You calculate the confidence interval estimate of the population slope by multiplying the t statistic by the standard error of the slope and then adding and subtracting this product to the sample slope.

WORKED-OUT PROBLEM 7   You want to determine the confidence interval estimate of the slope for the real estate study multiple regression model. From the spreadsheet results for WORKED-OUT PROBLEM 1, with 95% confidence, the lower limit for the slope of the property size with the asking price is $0.0164 thousand, and the upper limit is $0.0252 thousand. The confidence interval indicates that for each increase of 1 square foot of property size, the asking price is estimated to increase by at least $0.0164 thousand ($16.40) but less than $0.0252 thousand ($25.20), holding constant the age of the house.

With 95% confidence, the lower limit for the slope of the age of a home with the asking price is $-7.7731 thousand (-$7,773.10), and the upper limit is $-3.7612 thousand (-$3,761.20). The confidence interval indicates that for each increase of one year in the age of the house, the asking price is estimated to decrease by at least $3,761.20 but less than $7,773.10.

spreadsheet solution

Multiple Regression

Chapter 11 Multiple Regression contains the spreadsheet that WORKED-OUT PROBLEM 1 uses to create a multiple regression model for the real estate study data.

Best Practices

Use the **LINEST(*cell range of Y variable, cell range of X variable, True, True*)** function to calculate the b_0 and b_1 coefficients and standard errors, r^2, the standard error of the estimate, the F test statistic, the residual degrees of freedom, and *SSR* and *SSE*.

Use the **T.INV.2T(*1 – confidence level, residual degrees of freedom*)** function to calculate the critical value for the t test.

To calculate Significance F, the p-value for the F test, use the **F.DIST.RT(*F critical value, regression degrees of freedom, residual degrees of freedom*)** function.

To calculate the p-values in the spreadsheet ANOVA table, use the **T.DIST.2T(*absolute value of the t test statistic, residual degrees of freedom*)** function.

How-Tos

Experiment with this spreadsheet by changing the confidence level of significance in cell L8.

Examine the **Chapter 11 Multiple Regression** RESIDUALS spreadsheet, which does not appear in this chapter, for a model for calculating residuals.

Advanced Technique ATT6 in Appendix E describes using the Analysis ToolPak as a second way to perform a regression analysis.

Advanced Technique ADV6 in Appendix E explains more about how to use the LINEST function to calculate regression results.

One-Minute Summary

Multiple regression

- Use several independent variables to predict a dependent variable
- Net regression coefficients
- Coefficient of multiple determination
- Overall F test
- Residual analysis
- t test for the significance of each independent variable

Test Yourself
Short Answers

1. In a multiple regression model involving two independent variables, if b_1 is +3.0, it means that:
   a. the relationship between X_1 and Y is significant
   b. the estimated value of Y increases by 3 units for each increase of 1 unit of X_1, holding X_2 constant
   c. the estimated value of Y increases by 3 units for each increase of 1 unit of X_1, without regard to X_2
   d. the estimated value of Y is 3 when X_1 equals 0

2. The coefficient of multiple determination:
   a. measures the variation around the predicted regression equation
   b. measures the proportion of variation in Y that is explained by X_1 and X_2
   c. measures the proportion of variation in Y that is explained by X_1, holding X_2 constant
   d. will have the same sign as b_1

3. In a multiple regression model, the value of the coefficient of multiple determination:
   a. is between −1 and +1
   b. is between 0 and +1
   c. is between −1 and 0
   d. can be any number

Answer True or False:

4. The interpretation of the slope is different in a multiple linear regression model than in a simple linear regression model.

5. The interpretation of the Y intercept is different in a multiple linear regression model than in a simple linear regression model.

6. In a multiple regression model with two independent variables, the coefficient of multiple determination measures the proportion of variation in Y that is explained by X_1 and X_2.

7. The slopes in a multiple regression model are called net regression coefficients.

8. The coefficient of multiple determination is calculated by taking the ratio of the regression sum of squares over the total sum of squares (SSR/SST) and subtracting that value from 1.

9. You have just developed a multiple regression model in which the value of the coefficient of multiple determination is 0.35. To determine whether this indicates that the independent variables explain a significant portion of the variation in the dependent variable, you would perform an F test.

10. From the coefficient of multiple determination, you cannot detect the strength of the relationship between Y and any individual independent variable.

Answers to Test Yourself Short Answers

1. b
2. b
3. b
4. True
5. False

6. True
7. True
8. False
9. True
10. True

Problems

1. WORKED-OUT PROBLEM 1 creates a multiple regression model that predicts the asking price for homes in Glen Cove, New York, using the two independent variables property size and age of the home. Develop a multiple linear regression model that includes two additional independent variables: the number of bedrooms and the number of bathrooms.

Glen Cove

   a.   State the multiple regression equation.

   b.   Interpret the meaning of the slopes, b_1, b_2, b_3, and b_4, in this problem.

   c.   Explain why the regression coefficient, b_0, has no practical meaning in the context of this problem.

   d.   Predict the asking price for a house that has a property size of 10,004 square feet, 4 bedrooms, and 3.5 bathrooms and that is 65 years old.

e. Compute the coefficient of multiple determination, r^2, and interpret its meaning.

f. Perform a residual analysis on the results and determine the adequacy of the model.

g. At the 0.05 level of significance, determine whether a significant relationship exists between the asking price and the four indepen- dent variables property size, number of bedrooms, number of bath- rooms, and age of a home.

h. At the 0.05 level of significance, determine whether each independent variable makes a significant contribution to the regression model. On the basis of these results, indicate the independent variables to include in this model.

i. Construct a 95% confidence interval estimate of the population slope between asking price and property size, asking price and num- ber of bedrooms, asking price and number of bathrooms, and asking price and age.

j. Compare the predicted asking price in part (d) to the predicted asking price in WORKED-OUT PROBLEM 2. What might explain the difference?

2. Measuring the height of a California redwood tree is a very difficult under- taking because these trees grow to heights of more than 300 feet. People familiar with these trees understand that the height of a California redwood tree is related to other characteristics of the tree, including the diameter of the tree at the breast height of a person and the thickness of the bark of the tree. The following data represent the height (in feet), diameter at breast height of a person, and bark thickness for a sample of 21 California redwood trees:

Height	Diameter at Breast Height	Bark Thickness	Height	Diameter at Breast Height	Bark Thickness
122.0	20	1.1	164.0	40	2.3
193.5	36	2.8	203.3	52	2.0
166.5	18	2.0	174.0	30	2.5
82.0	10	1.2	159.0	22	3.0
133.5	21	2.0	205.0	42	2.6
156.0	29	1.4	223.5	45	4.3
172.5	51	1.8	195.0	54	4.0
81.0	11	1.1	232.5	39	2.2
148.0	26	2.5	190.5	36	3.5
113.0	12	1.5	100.0	8	1.4
84.0	13	1.4			

Redwood

a.   State the multiple regression equation that predicts the height of a tree, based on the tree's diameter at breast height and the thickness of the bark.

b.   Interpret the meaning of the slopes in this equation.

c.   Predict the height of a tree that has a breast diameter of 25 inches and a bark thickness of 2 inches.

d.   Interpret the meaning of the coefficient of multiple determination in this problem.

e.   Perform a residual analysis on the results and determine the adequacy of the model.

f.   Determine whether a significant relationship exists between the height of redwood trees and the two independent variables (breast diameter and the bark thickness) at the 0.05 level of significance.

g.   At the 0.05 level of significance, determine whether each independent variable makes a significant contribution to the regression model. Indicate the independent variables to include in this model.

h.   Construct a 95% confidence interval estimate of the population slope between the height of the redwood trees and breast diameter and between the height of redwood trees and the bark thickness.

Baseball

3. A baseball analytics specialist wants to determine which variables are important in predicting a team's wins in a given season. He has collected data related to wins, earned run average (ERA), and runs scored per game for a recent season. Develop a model to predict the winning percentage based on ERA and runs scored per game.

a.   State the multiple regression equation.

b.   Interpret the meaning of the slopes in this equation.

c.   Predict the winning percentage for a team that has an ERA of 3.75 and has runs scored per game of 4.00.

d.   Perform a residual analysis on the results and determine whether the regression assumptions are valid.

e.   Is there a significant relationship between winning percentage and the two independent variables (ERA and runs scored) at the 0.05 level of significance?

f.   Interpret the meaning of the coefficient of multiple determination.

g.   At the 0.05 level of significance, determine whether each independent variable makes a significant contribution to the regression model. Indicate the most appropriate regression model for this set of data.

h.   Construct 95% confidence interval estimates of the population slope between winning percentage and ERA and between the winning percentage and runs scored per game.

Domestic Beer

4. The file **Domestic Beer** contains the percentage of alcohol, number of calories per 12 ounces, and number of carbohydrates (in grams) per 12 ounces for 157 of the best-selling domestic beers in the United States. (Data extracted from "Find Out How Many Calories in Beer," https://www.beer100.com/beer-calories/).

Using this data, develop a multiple regression model to predict the calories in a domestic beer, based on the percentage of alcohol and the number of carbohydrates in the beer.

    a.   State the multiple regression equation.

    b.   Interpret the meaning of the slopes, b_1 and b_2.

    c.   Explain why the regression coefficient, b_0, has no practical meaning in the context of this problem.

    d.   Predict the calories for a beer that has an alcohol percentage of 5.0 and 10 grams of carbohydrates.

    e.   Compute the coefficient of multiple determination, r^2, and interpret its meaning.

    f.   Perform a residual analysis on your results and determine the adequacy of the fit of the model.

    g.   Determine whether a significant relationship exists between the calories and the alcohol percentage and the carbohydrates at the 0.05 level of significance.

    h.   At the 0.05 level of significance, determine whether each independent variable makes a significant contribution to the regression model. On the basis of these results, indicate the independent variables to include in this model.

    i.   Construct a 95% confidence interval estimate of the population slope between calories and alcohol percentage and calories and carbohydrates.

Answers to Problems

1. a.   Predicted asking price = 296.1752 + 0.0140 × property size + 46.3438 × number of bedrooms + 149.1666 × number of bathrooms − 3.1288 × age.

    b.   For a given property size, number of bedrooms, number of bathrooms, and age, each increase by 1 square foot in property size is estimated to result in an increase in asking price of $14.

        For a given property size, number of bathrooms, and age, each increase of one bedroom is estimated to result in an increase in asking price of $46.3438 thousand.

For a given property size, number of bedrooms, and age, each increase of one bathroom is estimated to result in an increase in asking price of $149.1666 thousand.

For a given property size, number of bedrooms, and number of bathrooms, each increase of one year in age is estimated to result in a decrease in asking price of $3.1288 thousand.

c. The interpretation of b_0 has no practical meaning because it would represent the estimated asking price of a new house that has no property size, no bedrooms, and no bathrooms.

d. Predicted asking price = 296.1752 + 0.0140(10,004) + 46.3438(4) + 149.1666 (3.5) −3.1288(65) = $939.9666 thousand.

e. $r^2 = 0.8208$; therefore, 82.08% of the variation in the asking price of a home can be explained by variation in property size, number of bedrooms, number of bathrooms, and age of the home.

f. There is no particular pattern in the residual plots, and the model appears to be adequate.

g. $F = 65.2615$; p-value $= 0.0000 < 0.05$. Reject H_0. There is evidence of a significant linear relationship between asking price and the four independent variables.

h. For property size: $t = 6.7064 > 2.0025$ or p-value $= 0.0000 < 0.05$. There is evidence that there is a significant relationship between property size and asking price, holding constant the other independent variables.

For number of bedrooms: $t = 1.7856 < 2.0025$ or p-value $= 0.0795 < 0.05$. There is insufficient evidence that there is a significant relationship between number of bedrooms and asking price, holding constant the other independent variables.

For number of bathrooms: $t = 4.817 > 2.0025$ or p-value $= 0.0000 < 0.05$. There is evidence that there is a significant relationship between number of bathrooms and asking price, holding constant the other independent variables.

For age: $t = -3.4699 < -2.0518$ or p-value $= 0.0010 < 0.05$. There is evidence that there is a significant relationship between age and asking price, holding constant the other independent variables.

i. $0.0098 < \beta_1 < 0.0181$; $-5.6274 < \beta_2 < 98.3150$; $87.4732 < \beta_3 < 210.8600$; $-4.9344 < \beta_4 < -1.3232$.

j. The predicted asking price in part (d) is more than in WORKED-OUT PROBLEM 2 because it considers the number of bedrooms and the number of bathrooms in addition to the property size and age.

2. a. Predicted height = 62.1411 + 2.0567 × diameter of the tree at breast height of a person (in inches) + 15.6418 × thickness of the bark (in inches).

   b. Holding constant the effects of the thickness of the bark, for each additional inch in the diameter of the tree at breast height of a person, the height of the tree is estimated to increase by 2.0567 feet. Holding constant the effects of the diameter of the tree at breast height of a person, for each additional inch in the thickness of the bark, the height of the tree is estimated to increase by 15.6418 feet.

   c. Predicted height = 62.1411 + 2.0567 (25) + 15.6418 (2) = 144.84 feet.

   d. $r^2 = 0.7858$; 78.58% of the total variation in the height of the tree can be explained by the variation in the diameter of the tree at breast height of a person and the thickness of the bark of the tree.

   e. The plot of the residuals against bark thickness indicates a potential pattern that might require the addition of curvilinear terms. One value appears to be an outlier in both plots.

   f. $F = 33.0134$ with 2 and 18 degrees of freedom. p-value = virtually $0 < 0.05$. Reject H_0. At least one of the independent variables is linearly related to the dependent variable.

   g. Breast height diameter: $t = 4.6448 > 2.1009$ or p-value = $0.0002 < 0.05$, Reject H_0. Breast height diameter makes a significant contribution to the regression model after bark thickness is included.

      Bark thickness: $t = 2.1882 > 2.1009$ or p-value = $0.0421 < 0.05$, Reject H_0. Bark thickness makes a significant contribution to the regression model after breast height diameter is included.

      Therefore, both breast height diameter and bark thickness should be included in the model.

   h. $1.1264 \leq \beta_1 \leq 2.9870$; $0.6238 \leq \beta_2 \leq 30.6598$.

3. a. Predicted winning percentage = 0.5893 − 0.1004 × ERA + 0.0769 × runs scored per game.

   b. For a given number of runs scored per game, for each increase of 1 in the ERA, the winning percentage is estimated to decrease by 0.1004. For a given ERA, for each increase of 1 in the number of runs scored per game, the winning percentage is estimated to increase by 0.0769.

   c. Predicted winning percentage = 0.5893 − 0.1004(3.75) + 0.0769(4) = 0.5204.

   d. There is no evidence of a pattern in the residual plot of ERA or runs scored per game.

   e. $F = MSR/MSE = 0.1059/0.0014 = 73.3829$; p-value = $0.0000 < 0.05$. Reject H_0. Evidence of a significant linear relationship exists between winning percentage and the two explanatory variables.

f. $r^2 = SSR/SST = 0.2118/0.2507 = 0.8446$; 84.46% of the variation in winning percentage can be explained by variation in ERA and runs scored per game.

g.   For X_1: $t_{STAT} = b_1 / S_{b_1} = -9.7256 > -2.0518$ and p-value $0.0000 < 0.05$; reject H_0. There is evidence that the variable X_1 contributes to a model already containing X_2. For X_2: $t_{STAT} = b_2 / S_{b_2} = 5.7724 > 2.0518$ and p-value $0.0000 < 0.05$; reject H_0. Both variables X_1 and X_2 should be included in the model.

h. $-0.1216 < \beta_1 < -0.0792$; $0.0495 < \beta_2 < 0.1042$.

4. a.   Predicted calories $= -5.1828 + 21.5146$ alcohol $+ 3.9387$ carbohydrates.

   b.   For a given carbohydrate, each increase of one unit in alcohol is estimated to result in an increase in calories of 21.5146. For a given alcohol percentage, each increase of one carbohydrate is estimated to result in an increase in calories of 3.9387.

   c.   The interpretation of b_0 has no practical meaning here because it would involve estimating calories for a beer that has 0 alcohol and 0 carbohydrates.

   d.   Predicted calories are $141.777 = -5.1828 + 21.5146\,(5) + 3.9387(10)$.

   e.   There appears to be no relationship in the plot of the residuals against the predicted values or against either of the two independent variables.

   f. $F = 2{,}275.4758$; p-value $= 0.0000 < 0.05$. Reject H_0. Evidence of a significant linear relationship exists between calories and alcohol and carbohydrates.

   g. $r^2 = 0.9673$. 96.73% of the variation in calories can be explained by variation in alcohol and variation in carbohydrates.

   h.   For alcohol: $t = 38.3303 > 0.0000$ or p-value $= 0.0008 < 0.05$, Reject H_0. There is evidence that alcohol contributes to a model that already contains carbohydrates.

         For carbohydrates: $t = 25.8068 < 0.0000$ or p-value $= 0.0000 < 0.05$. Reject H_0. There is evidence that carbohydrates contribute to a model that already contains alcohol. Both variables X_1 and X_2 should be included in the model.

   i. $20.4058 < \beta_1 < 22.6234$; $3.6372 < \beta_2 < 4.2402$.

References

1. Berenson, M. L., D. M. Levine, K. A. Szabat, and D. Stephan. *Basic Business Statistics: Concepts and Applications*, 15th edition. Hoboken, NJ: Pearson Education, 2023.

2. Kutner, M. H., C. Nachtsheim, J. Neter, and W. Li. *Applied Linear Statistical Models*, 5th edition. New York: McGraw-Hill-Irwin, 2005.

Introduction to Analytics

Analytics are methods that discover patterns in data with the aim of generating new or improved information from which better decisions can be made. Analytics may seem like a brand-new field, but many of the techniques are not new—even as the union of these techniques in a form that makes them easily accessible is.

Analytics has several antecedents, the most important one being statistics. In fact, many consider multiple regression, which Chapter 11 discusses, to be a foundational method in analytics. Therefore, learning statistics is a good first step towards learning analytics.

12.1 Basic Concepts

In any conversation or study about analytics, certain concepts or vocabulary terms are unavoidable. Because analytics combines techniques from several different fields, some concepts can be defined in several different ways, which can lead to fuzziness or confusion.

Data Science Versus (Applied) Analytics

CONCEPT   The two aspects of the same thing. Data science focuses more on providing the means to do analytics. Analytics uses those means to generate improved information and, ultimately, to assist in decision making. (This applied sense of analytics is how the title of this book uses the word.)

INTERPRETATION   The relationship between data science and analytics is similar to the relationship between the technical designers, developers, and implementors of software and those who *use* that software for some productive reason, such as searching the Internet or working with Microsoft Excel. Data science specialists manage the programs and data that provide the means to do analytics as well as help construct software solutions that make the tools of analytics accessible to non-technical staff.

Some use "analytics" to refer collectively to the two terms this section defines. Such usage can be confusing. If you discover an "Introduction to Analytics" course, you will not immediately know exactly what that course covers. The course might contain content similar to the content of this chapter and Chapter 13, or it might be a technical introduction to data science that includes using software such as R or Python to construct solutions.

In the applied sense, analytics often gains a modifier, as in "business analytics," "marketing analytics," or "health care analytics." A course title with these terms would typically include the use of preexisting software solutions to generate information—similar to the way this chapter uses Microsoft Excel.

Types of Analytics

CONCEPT   Analytics techniques and methods that are classified as being examples of descriptive analytics, predictive analytics, or prescriptive analytics.

INTERPRETATION   The three categories of analytics answer different questions, as the following table summarizes.

Analytics Category	Defining Question
Descriptive analytics	What has happened or has been happening?
Predictive analytics	What could happen?
Prescriptive analytics	What should happen?

Descriptive analytics extends descriptive statistics and predictive analytics extends inferential statistics and is the category that includes regression methods. The following separate concept entries explain more about the three categories.

Descriptive Analytics

CONCEPT Descriptive methods that summarize data, especially data that contains many variables or is a large collection of data.

EXAMPLES A visualization that summarizes multiple variables as a two-dimensional display. A summary table that enables a person to examine the underlying detail information. A real-time status display that summarizes complex operations, such as a display that presents the current status of every phase of a sports arena operation.

INTERPRETATION Descriptive analytics builds on descriptive statistics methods. Descriptive analytics methods produce results that are usually designed to be shared or viewed by more than one person and that may include the processing of streaming data that updates the results.

Individual descriptive results can be combined into one *dashboard* that serves as a master summary of some business process or objective and that can be constantly updated. Descriptive analytics results may incorporate an interactive *drill-down* capability which enables the exploration of decreasing levels of summarization.

Predictive Analytics

CONCEPT Methods that identify what is likely to occur in the near future and find relationships in data that might not be readily apparent using descriptive analytics.

EXAMPLES Predicting customer behavior in a retail context, detecting patterns of fraudulent financial transactions.

INTERPRETATION Predictive methods assign a value to target variables, classify items in a collection to target categories or classes, group or cluster items into natural groupings called clusters, or find items that tend to co-occur and specify the rules that govern their co-occurrence.

Many predictive analytics methods require software functionality not found in common tools such as Microsoft Excel. Chapter 13 discusses more fully explores this category of analytics.

Prescriptive Analytics

CONCEPT Methods that can suggest the best future decision making for specific case situations.

EXAMPLES Recommended daily optimal staffing levels for retailers that minimize labor costs but maximize sales; a system for a tour operator to set hotel pricing to maximize revenue while ensuring pricing consistency across all levels of accommodation offered.

INTERPRETATION   Prescriptive methods seek to improve organizational performance by optimizing decisions. Forerunners of these methods include the classical *optimization methods* used in management science and operations research. Prescriptive analytics leverage descriptive and predictive methods to create the basis for recommendation.

Prescriptive analytics builds on predictive analytics results in order to improve organizational performance. This makes this category of analytics an advanced topic that is beyond the scope of this book.

Drill Down

CONCEPT   The capability that enables the exploration of ever-decreasing levels of summarization.

EXAMPLES   A data visualization that enables a viewer to discover the data used to create the visualization. A summary table the enables a person to see various levels of decreasing summarization down to the original detail data.

INTERPRETATION   Drill down is an important feature in many descriptive analytics summaries. Drill down helps manage the complexity or detail of the data being summarized, showing only the level of complexity and detail necessary for individual decision makers.

Drill down also permits *data discovery*, the process by which decision makers can review data for patterns or exceptional or unusual values or outliers. Drill down also enables different decision makers who have different information needs to share (and discuss) one summary.

The dashboard concept in Section 12.2 includes a worked-out problem that shows an example of drill down.

Data Mining

CONCEPT   Using predictive analytics methods to discover previously unknown patterns in data, especially in large collections of data.

EXAMPLES   A streaming video service's recommendation engine for customers. A retailer discovering demographic information about its customers based on buying choices.

INTERPRETATION Data mining seeks to "mine" data resources to discover patterns, and the phrase derives from an analogy with natural resource mining. Because the phrase was originated by marketers, the exact meaning of data mining is sometimes fuzzy, with some using the term as a synonym for predictive analytics.

In the most frequent usage, data mining refers to doing predictive analytics combined with *feature selection*, a process that Chapter 13 discusses.

Machine Learning

CONCEPT Any computerized process that discovers or establishes a pattern in the data being analyzed.

INTERPRETATION As a term derived from computer science, machine learning is an alternative way of characterizing predictive and prescriptive methods. No machine actually "learns" anything. Instead, a software process attempts to create a model that reflects the data being analyzed. Regression methods do this, making them one of the simplest examples of *machine learning*.

In machine learning classification, methods are either supervised or unsupervised. These terms refer to whether the data contains *labeled examples* or not. Labeled examples provide hints about the data being analyzed.

For example, a visual processing method that uses picture data that contains a label that identifies the picture as a person or as a television or whatever is a supervised method. A visual processing method that uses unidentified pictures is an unsupervised method. Regression is a supervised method because the dependent variable (Y) values in the regression data serve as labels for the independent (X) values.

Unstructured Data

CONCEPT Collections of data that cannot be handled in traditional ways and that require advanced or innovative information processing techniques.

EXAMPLES Images, audio and video files, written language.

INTERPRETATION Structure in data means having a regular and repeating pattern. Data that can be made to fit a standardized form, such as a medical claim form, or the contents of a web page that asks for payment information, contains structured data. Any data that easily fits into a spreadsheet, such as the Excel data files for this book, is also structured data.

Unstructured data cannot be made to fit into a form easily. Think about collecting the contents of print-only web pages, all the social media posts about a topic, or the chapter contents of a book. If you were to put such data into a spreadsheet, you would need to put the *unstructured text* of each page, post, or chapter in its own cell in the same column. You would not be able to format the text into a regular and repeating pattern of two or more columns.

Furthermore, the entry in each cell would be of variable length. For example, a social media post might contain a few words or might contain a long essay and might also contain other types of unstructured data, such as images. In this variability, there would be no regular and recurring pattern to all social media posts.

Early computer systems required *structured data*, usually in the form of individual records composed of facts ("fields") and gathered together into a file. Such systems could not easily analyze unstructured text, images, video streams, and the like.

Today, some use the term *semi-structured data* to describe data that shares aspects of both structured and unstructured data. A social media post could be considered semi-structured because each post has some repeating elements. For example, an Instagram post would always have account identity information, a picture, and a caption, and the post would likely also have comments and hashtags, even though these things can vary greatly from one post to another.

Big Data

CONCEPT   High-volume, high-velocity, and/or high-variety collections of data that require innovative forms of information processing for enhanced insight and decision making.

EXAMPLES   A securities trading system that combines predictive models of price behavior with social and political trends to better estimate the value of an investment. A health maintenance organization's collection of patient, clinic, lab, hospital, and pharmacy data that is used with analytics to generate a prescriptive plan for delivering high-quality health care during an upcoming flu season.

INTERPRETATION   Big data was a phrase coined in 2001 by an industry analyst who was describing a future that has been somewhat realized, using analytics on a large scale. Today, mentions of big data imply *systems* that provide predictive information or prescriptive solutions for large organizations and systems that are equipped to handle a *variety* of structured, semi-structured, and unstructured data.

Twenty years ago, an early focus was on *volume*, the size of the data (the "big" in big data). A large volume of data is data with a large number of observations, a large number of variables, or that is composed of a large number of sets of data. Handling high-volume data required improvements in data management and data communications—improvements that have since been made and are ongoing today.

More recently, big data is more likely to refer to data that is generated (and used) at a very fast rate—that is, data that has a "big" *velocity*. A system that can handle such data can better reflect the current status of a process and provide the timeliest information for decision making.

Volume, velocity, and variety were all part of the industry analyst's original "3Vs" definition. In years since, others have added other *v* words, such as talking about the *veracity*, or trustworthiness, of data, or the *value* of the data to a decision-making process. Regardless of the number of *v* words used, the term *big data* today implies large-scale, organizational systems of the type listed under Examples for this concept.

12.2 Descriptive Analytics

Descriptive analytics summarize data, especially data that is *big data*. Descriptive analytics extends descriptive statistics by presenting more data in one tabular or visual summary and by offering interactivity such as drill down that enables further exploration of the summarized data.

Dashboard

CONCEPT   A visual display that summarizes the most important variables needed for decision making or achieving a business objective.

EXAMPLES   A display that presents the financial status of a business for senior managers; a display that summarizes the sales of a national retail chain in a variety of ways, including by store or by item category sold; a display that monitors the activities of all employees in a call center; a public video wall that displays information such as the NASDAQ MarketSite video wall in Times Square, New York, which summarizes NASDAQ securities exchange transactions.

INTERPRETATION   Dashboards create opportunities for group discussion by enabling sharing of summarized information. For several decades, people imagined and discussed information systems that would put information at the "fingertips" of decision makers; dashboards seek to do that for the most important information in a decision-making process.

Dashboards can include the summary tables and visuals that Chapter 2 discusses, as the Figure 12.1 amusement park store sales dashboard on page 266 illustrates. Dashboards can also contain the newer and more novel visualizations that Section 12.3 discusses. While dashboards are useful for sets of data of any size, they are most often used to summarize large collections of data, including sets that are big data.

FIGURE 12.1

A sales dashboard that contains a bar chart, data table, and a line chart

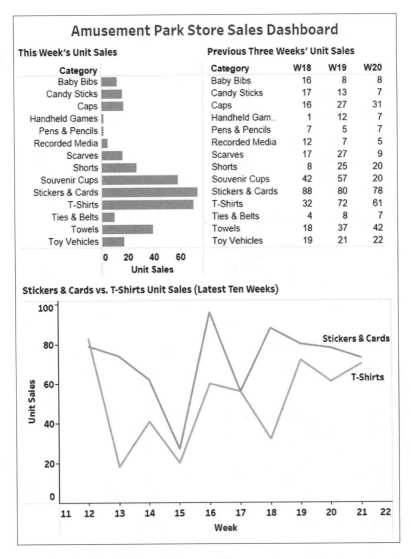

Multidimensional Contingency Table

CONCEPT   A tabular summary of three or more variables, typically two or three categorical variables and one numerical variable.

EXAMPLES   An income tax table that presents tax due amounts classified by filing status and level of adjusted gross income; the runs/hits/errors box of a baseball box score; a multi-year accounting or financial statement; an Excel PivotTable that summarizes the 10-year returns of mutual funds by fund type, market capitalization, and level of risk.

INTERPRETATION A multidimensional contingency table (MCT) is an extension of the two-way table concept that includes additional variables. MCTs are often implemented as *pivot tables*, which include drill-down features and the ability to change ("pivot") the statistics being displayed.

Some use the term pivot table as a synonym for a *MCT*, but, strictly speaking, a pivot table can implement a summary or a two-way table as well as a multidimensional contingency table.

WORKED-OUT PROBLEM 1 You have been asked to analyze a sample of mutual funds that has been summarized in a multidimensional contingency table. The MCT summarizes the mean 10-year return percentages of the mutual funds by fund type (growth or value), market capitalization (small, midcap, or large), and level of risk (low, average, or high). The MCT is to be implemented using an Excel *PivotTable*.

The Figure 12.2 fully summarized PivotTable shows that value funds with low or high risk have a higher mean 10-year return percentage than the growth funds with those same risk levels. Drilling down, the expanded table (Figure 12.2 right) reveals a more complicated pattern. This more detailed table shows that mutual funds with large market capitalizations are the poorest performers and that growth funds that have large market capitalizations significantly depress the mean for all growth funds.

The drill down also reveals other relationships; for example, it shows that value funds with small market capitalizations are the funds that perform the best. The blank in the cell for large value funds with high risk reveals that, in this sample, there are no value funds with a high level of risk and a large market capitalization.

FIGURE 12.2

Excel PivotTables for a sample of mutual funds fully summarized (left) and expanded (right)

	A	B	C	D	E
1	Contingency Table of Fund Type, Market Cap, and Risk, showing the mean ten-year return percentage				
2					
3	Mean 10-Yr Return	RISK ▾			
4	TYPE ▾	Low	Average	High	Grand Total
5	⊞ Growth	16.12	17.07	16.72	16.73
6	⊞ Value	17.14	16.71	18.87	17.47
7	Grand Total	16.50	16.99	17.48	16.95

	A	B	C	D	E
1	Contingency Table of Fund Type, Market Cap, and Risk, showing the mean ten-year return percentage				
2					
3	Mean 10-Yr Return	RISK ▾			
4	TYPE ▾	Low	Average	High	Grand Total
5	⊟ Growth	16.12	17.07	16.72	16.73
6	Large	15.69	15.65	13.26	15.48
7	Mid-Cap	17.62	18.04	17.77	17.92
8	Small	17.38	18.15	17.30	17.65
9	⊟ Value	17.14	16.71	18.87	17.47
10	Large	16.52	16.13		16.34
11	Mid-Cap	18.62	18.27	17.52	18.01
12	Small	22.77	20.12	19.58	19.94
13	Grand Total	16.50	16.99	17.48	16.95

A further drill down of the midcap funds with low risk (see Figure 12.3) uncovers another difference between the growth and value funds. For midcap value funds with low risk, the funds with the largest numbers of assets have the lowest expense ratios. In contrast, for midcap growth funds with low risk (see Figure 12.3 bottom), the lowest expense ratios are associated with funds with modest asset sizes.

FIGURE 12.3

Further drill down of Excel PivotTables examining midcap funds with low risk (top) and midcap growth funds with low risk (bottom)

	A	B	C	D	E	F	G	H	I
1	Fund Number	Market Cap	Type	Assets	Turnover Ratio	Beta	Risk	10-Yr Return	Expense Ratio
2	RF241	Mid-Cap	Value	44.7	21.0	1.02	Low	17.43	1.27
3	**RF239**	Mid-Cap	Value	**1452.0**	43.0	0.93	Low	20.2	**1.01**
4	RF238	Mid-Cap	Value	196.9	68.1	0.76	Low	18.68	1.70
5	**RF235**	Mid-Cap	Value	**1546.1**	41.0	0.99	Low	20.9	**1.15**
6	RF231	Mid-Cap	Value	37.1	82.0	0.86	Low	15.88	1.76

	A	B	C	D	E	F	G	H	I
1	Fund Number	Market Cap	Type	Assets	Turnover Ratio	Beta	Risk	10-Yr Return	Expense Ratio
2	RF222	Mid-Cap	Growth	70.5	205.0	0.59	Low	14.07	1.81
3	RF221	Mid-Cap	Growth	150.6	150.0	0.82	Low	19.82	1.30
4	RF217	Mid-Cap	Growth	135.4	7.0	0.76	Low	12.53	1.31
5	RF216	Mid-Cap	Growth	9.1	246.0	0.95	Low	14.68	1.17
6	**RF208**	**Mid-Cap**	**Growth**	**110.6**	**27.9**	**0.88**	**Low**	**16.88**	**0.99**
7	RF207	Mid-Cap	Growth	3507.4	18.0	0.93	Low	21.76	1.14
8	RF203	Mid-Cap	Growth	174.0	12.0	0.92	Low	17.99	1.25
9	RF202	Mid-Cap	Growth	61.8	18.0	0.98	Low	18.18	1.46
10	RF200	Mid-Cap	Growth	287.6	16.0	0.95	Low	18.73	1.21
11	RF190	Mid-Cap	Growth	27.9	159.0	0.79	Low	20.06	2.00
12	RF188	Mid-Cap	Growth	319.5	7.0	0.96	Low	19.3	1.23
13	**RF184**	Mid-Cap	Growth	**95.4**	35.0	0.92	Low	17.38	**1.04**

spreadsheet solution

Multidimensional Contingency Tables

Chapter 12 PivotTable contains the PivotTables and drill-down detail worksheets of WORKED-OUT PROBLEM 1 as separate worksheets.

Best Practices

Use the Excel PivotTable feature to create drill-down-enabled multidimensional contingency tables.

How-Tos

Experiment by double-clicking the various cells of one of the PivotTables in **Chapter 12 PivotTable** to produce additional worksheets that reveal the details about a particular group of mutual funds.

12.3 Typical Descriptive Analytics Visualizations

Sparklines

CONCEPT   A tabular set of visualizations that summarize a set of measurements taken over time as small, compact graphs that appear as part of a table or written passage.

EXAMPLES   A table of sparklines that shows changes in stock prices for a set of stocks; a table of sparklines that compares drug therapy responses in a set of patients; a table of sparklines that presents changes in sales over time for a set of competing manufacturers.

INTERPRETATION   Sparklines are an alternative to line charts that visually summarize *time-series* data, measurements taken over time. Sparklines are a better choice than line charts when sets of measurements have similar values and when plotting each set as a line graph would cause too much overlap. Sparklines are also more helpful than line charts when examining the changes in values is more important than examining the values themselves.

WORKED-OUT PROBLEM 2   The following table reports last year's changes in sales of four automobile manufacturers.

Company	Yearly Change
Alpha	Up 3.0%
Bravo	Up 33.2%
Charlie	Up 6.3%
Delta	Up 11.8%

You want to better understand what these percentage changes mean in a historical sense, so you create the Figure 12.4 table of sparklines that summarizes sales for the four manufacturers for the past four years. For each company in this table, a blue point represents the year with the greatest sales and a red point represents the year with the fewest sales.

FIGURE 12.4

Sparklines table that summarizes sales for the four manufacturers for the past four years

Company	Four-Year Sparklines
Alpha	
Bravo	
Charlie	
Delta	

The sparklines show that during year three, the previous year, all four companies saw sales drop to the lowest point in the four-year period. By providing a historical context, the sparklines reveal that company Bravo's exceptional yearly change reflects a severe drop in sales in the previous year and that company Delta's sales have greatly declined over the four years. The sparklines also show that company Charlie's sales have been least subject to fluctuations during the four-year period.

spreadsheet solution

Sparklines

Chapter 12 Sparklines contains the sparklines table that WORKED-OUT PROBLEM 2 uses to analyze four-year trends in the sales of automobile manufacturers.

Best Practices

In the Excel **Insert tab**, select **Line Sparkline** in the **Sparklines group** to create a sparklines table using the line sparklines that this section discusses.

To ensure that all sparklines in a sparkline table share the same vertical scale, select **Sparkline → Axis** and click **Same for All Sparklines** under **Vertical Axis Minimum Value Options**. Then select **Sparkline → Axis** a second time and click **Same for All Sparklines** under **Vertical Axis Maximum Value Options**.

How-Tos

Experiment with the sparklines table by changing the yearly sales data in columns B through E.

Treemap

CONCEPT   A chart that visualizes the comparison of two or more variables using the sizes and colors of rectangles to represent values.

EXAMPLES   A treemap that shows relative sales of models from several automobile manufacturers; a treemap that shows statewide voting patterns by county as blocks sized by the number of voters in a county; a treemap that compares the relative size of assets for a group of mutual funds, classified by fund type.

INTERPRETATION   A treemap uses rectangular areas of various sizes to illustrate relative differences of some measure. Treemaps use color to either indicate categories of a categorical variable (Excel) or to summarize a range of values of a numerical variable (Tableau, another visualization tool).

Treemaps trade precision for quick impressions. In decision-making situations that concern managing the status of some activity, quick impressions can be more useful than precision *if* the variables that the treemap visualizes are thoughtfully chosen.

WORKED-OUT PROBLEM 3   You have been asked to help prepare a presentation that examines a sample of small market capitalization mutual funds that have a low level of risk. You create the Figure 12.5 Excel treemap that compares the assets of growth and value funds in the sample. The treemap reveals that the largest fund in terms of assets is a growth funds and that growth fund show a wide range of asset values.

FIGURE 12.5

Excel treemap for mutual funds presentation

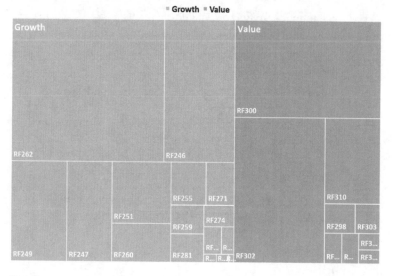

To offer the presenter an option, you then use Tableau to prepare the Figure 12.6 on page 272. The Figure 12.6 left treemap is similar to the Excel treemap, while the Figure 12.6 right treemap uses the values of the 10-year return variable to color the asset rectangles various shades of gray.

The second Tableau treemap reveals that the highest 10-year returns are not associated with the largest (by assets) funds. This treemap also shows that growth funds include a larger proportion of funds that have lower 10-year returns.

FIGURE 12.6

Tableau treemap for mutual funds presentation using color for fund type (left) or color for 10-year return percentages (right)

Assets of Small Market Cap Funds with Low Risk, by Fund Type

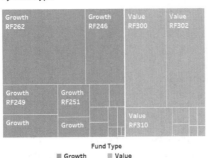

Alternative Treemap with Assets (size) and 10-Year Return (color)

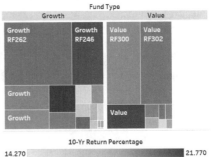

spreadsheet solution

Treemaps

Chapter 12 Treemap contains the treemap that WORKED-OUT PROBLEM 3 uses to analyze a sample of mutual funds.

Best Practices

Place the categorical data and the numerical data together in a series of consecutive columns with the categorical data first *before* selecting **Insert Hierarchy Chart → Treemap** from the Excel **Insert tab** to create a new treemap.

How-Tos

Experiment with the treemap by changing the data in the Small-LowDATA worksheet and then viewing the updated treemap.

Bullet Graph

CONCEPT   A visualization of a measurement as a bar chart plotted against a visualization of several numerical ranges that helps characterize the value of the measurement. Typically, a bullet graph contains a set of bar charts, each representing one category from a set of related categories.

EXAMPLES   A bullet graph that visualizes the key performance indicators for a business; a bullet graph that tracks student progress through a series of programmed assignments; a bullet graph that compares actual sales to sales targets for various merchandise categories.

INTERPRETATION   Bullet graphs enhance bar charts by providing context for the evaluation of the measurement value being visualized by a bar. A bullet graph may include a target line or marker that represents a desired numerical goal or limit. Stephen Few, the developer of bullet graphs, sees bullet graphs as being well suited for dashboards because they provide a compact way of comparing related measurements.

The visualized numerical ranges represent categories such as poor, acceptable, and excellent that are typically used as measures of performance. Bullet graphs can be oriented horizontally or vertically, and they may include special values other than targets or goals.

The bullet graph is not a chart type in current versions of Excel but can be created using a custom combination chart that uses five *data series* to produce a bullet graph. The bullet graph is a visualization type in other programs, including all versions of Tableau.

WORKED-OUT PROBLEM 4   You have been asked to characterize the online unit sales for several categories of online merchandise sold by an amusement park. You create the Figure 12.7 bullet graph shown below, in which the black bars represent current unit sales, and the blue lines represent unit sales targets. The darkest shading represents the weak-selling range, and the lighter shading represents an acceptable, but not excellent, selling range.

FIGURE 12.7

Bullet graph for online merchandise unit sales by category

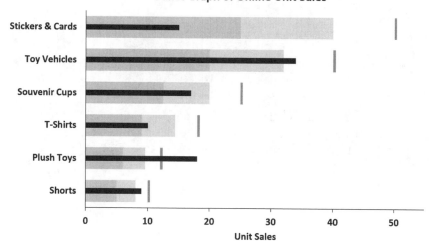

The bullet graph reveals that toy vehicles, plush toys, and shorts are doing very well, and the sales of plush toys is even exceeding the sales target. In contrast, online sales of stickers and cards are very weak, and the sales of souvenir cups and t-shirts are only acceptable. These observations would be helpful for merchandising managers making decisions about which categories and amounts of merchandise to stock for future sales.

spreadsheet solution

Bullet Graphs

Chapter 12 Bullet Graph contains the spreadsheet that WORKED-OUT PROBLEM 4 uses to analyze online sales for several categories of merchandise. Experiment with this chart by changing the target and actual values in columns B and C of the BulletData worksheet.

Best Practices

Label or otherwise explain target lines or markers that a bullet graph uses.

Minimize the use of color for shading, especially colors that could be distracting or colors that could be confusing for those with certain visual disabilities.

How-Tos

Experiment with this chart by changing the type, target, or actual values in columns A, B, and C of the BulletData worksheet in **Chapter 12 Bullet Graph**.

Advanced Technique ADV8 in Appendix E discusses how to modify the five data series of the bullet graph in **Chapter 12 Bullet Graph** for other problems.

One-Minute Summary

Basic concepts

- Analytics: descriptive, predictive, and prescriptive
- Drill down
- Data mining
- Machine learning
  - Unstructured data and big data

Descriptive analytics

- Dashboard
  - Multidimensional contingency table
- Choosing the correct descriptive analytics visualization:
  - To provide a historical context for *time-series* data, use sparklines.
  - To provide a preliminary comparison among numerical variables divided into categorical groups, use a treemap.
  - To present a single measurement or a set of related measurements and provide a reference by which the values can be evaluated and, optionally, to compare to a reference value, use a bullet graph.

Test Yourself

1. Methods that focus on discovering data patterns with the aim of generating new or improved information from which better decisions can be made are called:
   a. confidence intervals
   b. regression
   c. analysis of variance
   d. analytics

2. Methods that summarize data, especially data sets that contain many variables or large collections of data, are called:
   a. descriptive analytics
   b. predictive analytics
   c. prescriptive analytics
   d. all of the above

3. Methods that identify what is likely to occur in the near future and find relationships in data that might not be readily apparent are called:

    a.   descriptive analytics

    b.   predictive analytics

    c.   prescriptive analytics

    d.   all of the above

4. Methods that can suggest the best future decision making for specific case situations are called:

    a.   descriptive analytics

    b.   predictive analytics

    c.   prescriptive analytics

    d.   all of the above

5. Big data:

    a.   implies the use of equipment to handle a variety of data

    b.   is a term that was coined to describe a future that has been somewhat realized today

    c.   is data that is likely to be generated at a very fast rate

    d.   All of the above are correct.

6. The visualization that summarizes the most important variables needed for decision making or achieving a business objective is a:

    a.   dashboard

    b.   sparklines

    c.   treemap

    d.   bullet graph

7. The visualization that summarizes measurements taken over time in small, compact graphs designed to appear as part of a table or written passage is a:

    a.   dashboard

    b.   sparklines

    c.   treemap

    d.   bullet graph

8. The visualization that helps compare multiple variables using the size and color of rectangles to represent values is a:

    a.   dashboard

    b.   sparklines

    c.   treemap

    d.   bullet graph

9. An enhancement of a bar chart that provides a context for evaluating the measurement being visualized as a bar is a:
   a. dashboard
   b. sparklines
   c. treemap
   d. bullet graph

10. Drill down:
    a. can show various levels of summarization
    b. helps manage the complexity of a set of data
    c. is an important feature of dashboards
    d. All of the above are correct.

Answer True or False:

11. Analytics focuses on discovering patterns in or generating hypotheses about data already collected.

12. Descriptive analytics methods summarize large collections of data and present constantly updated status information.

13. Predictive analytics methods identify what is likely to occur in the near future and find relationships in data that might not be readily apparent.

14. Prescriptive analytics methods identify what is likely to occur in the near future and find relationships in data that might not be readily apparent.

15. Data mining can involve discovering patterns in data stored in the data resources of an organization.

16. A set of eight facts about 25 mutual funds arranged as 8 columns and 25 rows of a spreadsheet is an example of unstructured data.

17. Analytics requires the use of big data and cannot be applied to smaller sets of data.

18. Currently, Microsoft Excel implements all descriptive analytics methods that this chapter discusses.

19. A dashboard summarizes a set of variables needed for decision making.

20. Multidimensional contingency tables enable you to break down a variable according to the values of another variable in order to see patterns that might not be apparent in the original summary.

21. Drilling down always reveals more details about the data being summarized.

22. Sparklines display measurements for a single point of time.

23. Color in an Excel treemap can represent the values of a numerical variable or a categorical variable.

24. The shading in a bullet graph visualizes the same variable that the bars visualize.

Answers to Test Yourself

1. d	13. True
2. a	14. False
3. b	15. True
4. c	16. False
5. d	17. False
6. a	18. False
7. b	19. True
8. c	20. True
9. d	21. True
10. d	22. False
11. True	23. False
12. True	24. False

References

1. Few, Stephen. *Information Dashboard Design*, 2nd edition. Burlingame, CA: Analytics Press, 2013.

2. Levin, Yuri. "Prescriptive Analytics: The Key to Improving Performance," SAS Knowledge Exchange, https://bit.ly/1pa4H6L.

3. Levine, D. M., D. F. Stephan, and K. A. Szabat. *Statistics for Managers Using Microsoft Excel*, 9th edition. Hoboken, NJ: Pearson Education, 2021.

4. Perceptual Edge. *Bullet Graph Design Specification*, https://bit.ly/1pal7f9.

5. Tufte, Edward. *Beautiful Evidence*. Cheshire, CT: Graphics Press, 2006.

Predictive Analytics

Learning about predictive analytics is more challenging than learning about the other concepts that this book discusses. To learn about predictive analytics, you must first be aware of the concepts discussed in previous chapters.

13.1 Predictive Analytics Methods

Predictive analytics answers "What could happen?" types of questions. Methods identify what is likely to occur in the near future and find relationships in data that might not be readily apparent using descriptive analytics.

Types of Predictive Analytics

CONCEPT   Target-based, classification, clustering, and association methods.

EXAMPLES   Multiple regression (a target-based method); classification tree (a classification method); k-means clustering (a clustering method); market basket analysis (an association method).

INTERPRETATION   Predictive analytics methods use one of the four techniques that Table 13.1 on the next page summarizes.

TABLE 13.1
Predictive analytics methods

Predictive Method Type	Goal of Method
Target-based	Predicting the value of one variable based on the value of other variables
Classification	Grouping items into two or more groups or classes
Clustering	Grouping similar items together
Association	Discovering relationships among specific values of multiple variables

Target-Based Methods

CONCEPT   Methods that predict the value of one variable based on the value of other variables.

INTERPRETATION   Multiple regression, which Chapter 11 discusses, is the most widely used target-based method. In multiple regression, the values of independent variables predict the value of the dependent variable. In terms of machine learning (which Chapter 12 defines), target-based methods use *labeled examples* and therefore are examples of *supervised learning*.

Classification Methods

CONCEPT   Methods that group items into two or more groups or classes.

INTERPRETATION   Classification methods seek to subdivide the data being analyzed. Some methods, such as the tree induction methods that this chapter discusses, aim to make multiple subdivisions to create groupings that are as homogenous as possible. Traditional classification methods, such as tree induction, are also examples of supervised learning, although newer methods that use unsupervised learning exist as well.

Clustering Methods

CONCEPT   Methods that discover groupings, called *clusters*, among items in a set of data.

INTERPRETATION   Clustering methods look for how alike items are. Determinations about how alike items are based on a measure of similarity, or dissimilarity, among items. Individual methods differ in how they calculate or use the distance measure.

Traditional clustering methods are examples of unsupervised learning. As such, the clusters found may be based on abstract data relationships that do not easily map to real-world phenomena.

Association Methods

CONCEPT   Methods that find sets of variable values that tend to co-occur or the rules that specify the co-occurrence of the values of two or more variables.

INTERPRETATION   Association methods use one of several techniques for finding co-occurrences of pairs or sets of variable values. Association methods create models that predict how likely a data relationship is. That makes the if-then rules probabilistic, unlike a rule of logic, such as "if a, then b," which, if true, always holds.

As examples of unsupervised methods, association methods can sometimes find unexpected or unanticipated relationships among things. In one famous retail case, purchasing beer at a certain time of day made it likely that the customer was also purchasing diapers.

13.2 **More About Predictive Models**

All predictive models enable one to discover patterns in the set of data being analyzed. Chapters 10 and 11 discuss regression models, which can be explained as mathematical statements that express relationships among variables being analyzed. In Chapter 10, the statement "Predicted value of opening weekend box office gross = 3.9309 + 0.8481 × Trailer views" expresses the regression model for the movie distribution analysis. Models created using classification, clustering, or association methods are not so easily expressed.

In common usage, a *model* is something that is a simplified representation of an object in the real world. A model toy car or an architect's 3D rendering of a building are scaled-down copies of tangible things. A statistical or analytical model also is a scaled-down representation but of something that is more abstract—the mathematical relationships among variables.

As simplifications, all models omit details of the thing they are modeling to some degree. A child's toy model car may leave out many details, including details of the car's interior and what is in the engine compartment. An adult-collectible toy model car would leave out much less and might reproduce the details of the interior and even have working parts that show how the engine delivers power to the wheels.

The adult-collectible model car would be a much better representation of a real-world car, but even such a model would not tell someone how the car drives on bumpy or potholed streets or provide an understanding of the car's acceleration or its fuel economy or range. A person who chose to buy or lease a car based solely on looking at adult collectible models might be disappointed upon discovering the flaws that the collectible model car did not or could not *model*.

Likewise, in answering "What could happen?" questions, predictive analytics methods may create models that leave out details that may later turn out to be significant omissions. But deciding what to include in a model can be hard to do, and different people might make different judgments. The previous paragraph concludes that a person "might be disappointed" and not "will be disappointed" because, perhaps, there is a car buyer who only cares about the look of real-world cars and is uninterested in their performance characteristics. For that person, the collectible model might be a sufficient basis for making a choice.

Limitations of Models

Predictive models supply insights that can lead to making better-informed decisions. Predictive models are not designed to replace human decision making and cannot replace such decision making because all models leave out details that may prove to be important. In one well-known example, warehouse workers for a large online retailer were summarily fired when a predictive model failed to make distinctions between chronic absenteeism and absences caused by serious illness.

Model Validation

CONCEPT   The process that ensures that a statistical or analytical model properly reflects the mathematical relationship being modeled and does so without violating any prerequisite assumption that the model needs to make.

INTERPRETATION   Tangible models such as toy cars can be directly examined for flaws. A Ford Mustang model car that has the grille work and logo of a Toyota Camry is rather obviously an incorrect model. Similarly, a Ford Mustang model car with only three wheels or with six doors is a defective model. A statistical or analytical model that uses the wrong data or violates necessary assumptions cannot be so easily discerned.

A simple linear regression model can be inspected by using a scatter plot that visualizes the model and illustrates the degree of correlation between the values of the independent variable and the values of the dependent variable. However, other types of predictive models are not so easily inspected. Such models need to be validated.

In learning about predictive analytics, many rush to use a predictive model without learning about the need for model validation. Understanding that need is perhaps the most important concept one can learn about analytics.

Cross-validation

CONCEPT   The validation process for predictive analytics models that use supervised learning.

INTERPRETATION   In cross-validation, the data to be analyzed are divided, and the divided data are analyzed separately. In the simplest case, one division creates a *training set* and a *test set*, in which the training set is used to create a method, and the test set is used to evaluate how well that model makes predictions.

In leave-one-out cross-validation, the data set is used repeatedly, each time omitting a different occurrence, to create a series of models that are then averaged together. For larger data sets, the k-fold technique divides the data into k parts (or "folds") and creates a series of models, each one based on data in which a different part has been omitted.

Some techniques use three-part data in which a training set and a *validation set* help form the predictive model, and the test set provides the evaluation of the finalized model.

Unsupervised Learning and Models

As Chapter 12 first discusses, all predictive analytics methods are examples of machine learning, and machine learning methods are either supervised or unsupervised. In supervised methods, the data to be analyzed include labels that tell the method something about the data. For example, in regression, a supervised method, the dependent variable value "labels" the other data and serves as an exemplar for what the regression model should predict as the value for the dependent variable.

Unsupervised methods do not have exemplars. For example, traditional cluster analysis data is unlabeled in the sense that each occurrence of data does not come with a label that identifies the cluster to which it belongs. Therefore, unsupervised methods may develop models that cannot be easily explained or mapped to real-world knowledge.

Without exemplars, unsupervised methods may "learn" the wrong relationships and develop models that make incorrect predictions. For example, an early facial recognition model consistently identified people with dark complexions as being great apes. An investigation revealed that the picture data that the model had used included very few pictures of people with dark complexions. The model had failed to learn that people can have dark complexions.

Currently, data scientists are devising *semi-supervised* machine learning, in which some of the data are *hints*, labeled examples that help guide an unsupervised method build a model. These methods have not been fully commercialized but are expected to be included in readily available software in the future.

13.3 **Tree Induction**

CONCEPT   Methods that classify data into groups based on the values of independent variables.

INTERPRETATION   Tree induction methods create models that can be visualized as tree diagrams that branch into forks at each level, or *node*. Tree analysis seeks to determine which values of a specific independent variable are useful in predicting the dependent variable. To construct a tree, a tree method chooses the one independent variable that provides the best split of the data at each node in the tree, starting with the root. These methods must also employ rules for deciding when a branch cannot be split anymore.

Classification and regression tree methods are not affected by the distribution of the variables that make up the data. Typically, trees are developed through several levels of nodes until either no further gain in the fit occurs or the splitting has continued as far as possible. After splitting, some methods *prune* the tree, deleting splits that do not enhance the final analysis.

Classification Tree

CONCEPT   The tree model that results from using a categorical dependent variable.

WORKED-OUT PROBLEM 1   You want to develop a classification tree to predict the probability that a customer of a wholesale distributor will accept an offer to contract for enhanced services. You collect a sample of 400 shipments and note the invoice amounts for six categories of goods (beverage, bakery, grocery, frozen, organic, and paper goods) as well as whether the buyer was using the basic or enhanced services. The following table shows the data for the first five shipments selected for the sample.

Analysis of Sales

Sales	Beverage	Bakery	Grocery	Frozen	Organic	Paper
Basic	5,567	871	2,010	3,383	375	569
Basic	31,276	1,917	4,469	9,408	2,381	4,334
Enhanced	6,353	8,808	7,684	2,405	3,516	7,844
Basic	13,265	1,196	4,221	6,404	507	1,788
Enhanced	22,615	5,410	7,198	3,915	1,777	5,185

You divide the data into a training set and a test set by randomly selecting 20% of the shipments as the test set. Figure 13.1 on the next page shows the resulting validated tree. This tree first splits customers on the basis of the amount spent on organic items and then makes a second split based on the amount spent on bakery items in one of the *subtrees*. (Figure 13.1 and the other illustrations in this chapter were created by the JMP software which reference 1 discusses.)

FIGURE 13.1

Chi-square
test results for
the web page
A/B test study

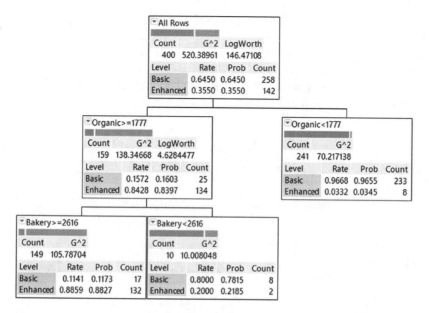

The Figure 13.1 classification tree reveals that customers using enhanced services tend to spend more on organic and bakery items. Future customers who spend more than $1,777 on organic items and $2,616 on bakery items might be targeted as prospects that would contract for enhanced services. In contrast, future customers who spend less than $1,777 on organic items would not be good prospects for enhanced services.

Regression Tree

CONCEPT   The tree model that results from using a numerical dependent variable.

WORKED-OUT PROBLEM 2   You want to better understand how engine attributes affected the miles per gallon (MPG) rating of vintage automobiles, and you suspect that engine weight may be the most important factor. You collect data on the MPG rating, engine horsepower (hp), engine weight (in pounds), acceleration (in seconds), and engine displacement (in liters) for 392 vintage 1970s automobiles. The following table shows the data for the first five automobiles selected for this sample.

**Vintage
Autos**

MPG	Horsepower	Weight	Acceleration	Displacement
18	130	3,504	12.0	5.03
15	165	3,693	11.5	5.74
18	150	3,436	11.0	5.21
16	150	3,433	12.0	4.98
17	140	3,449	10.5	4.95

You divide the data into a training and a test set by randomly selecting 20% of the shipments as the test set. Figure 13.2 shows the resulting validated tree. This tree first splits the mean MPG rating of vintage cars on whether the engine displacement is less than 3.24 liters. In each subtree, the tree splits the data on horsepower values. Only at the third (left subtree) or fourth (right subtree) level does the tree split the mean MPG ratings based on weight.

FIGURE 13.2

Chi-square test results for the web page A/B test study

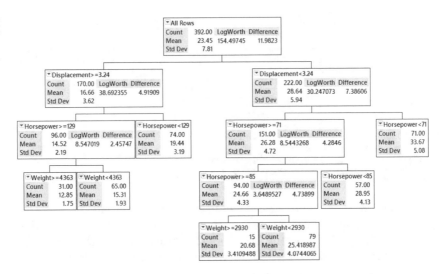

The Figure 13.2 regression tree reveals that, for vintage autos, displacement was much more significant than engine weight in affecting the MPG rating. For cars with a displacement of less than 3.24 liters, the model predicts a mean MPG of 28.64 MPG, and for cars with a displacement of 3.24 liters or more, the model predicts a mean MPG of only 16.66 MPG.

In this model, the subgroup with the highest mean MPG rating (33.67 MPG) is vintage autos with displacement less than 3.24 liters and horsepower less than 71 hp. In contrast, the subgroup with the lowest mean MPG rating (12.85 MPG) is vintage autos with displacements of 3.24 liters or more, horsepower of 129 hp or more, and weight of 4,363 pounds or more.

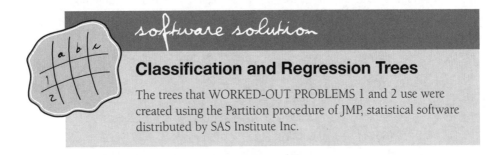

software solution

Classification and Regression Trees

The trees that WORKED-OUT PROBLEMS 1 and 2 use were created using the Partition procedure of JMP, statistical software distributed by SAS Institute Inc.

> **How-Tos**
>
> Although discussing JMP is beyond the scope of this book, readers interested in exploring tree induction using JMP can download a 30-day trial version of JMP at **www.jmp.com**. Also available at that website are tutorials and webinars about JMP and full documentation for using the Partition procedure.

13.4 Clustering

CONCEPT   Methods that assign individual occurrences to one of a set of groups, or *clusters*, such that occurrences in each cluster are more alike among themselves than they are like occurrences in other clusters.

INTERPRETATION   How clustering methods determine which occurrences are more alike than not varies. Hierarchical clustering and *k*-means clustering illustrate the two most-used techniques to determine the *similarity* among occurrences, which, in common usage, are called *items* when discussing clustering.

In hierarchical clustering, each item starts by itself, in its own cluster. The variable values of each item are examined, and the two items with variable values that are most similar are combined into one cluster. This inspection and combination of items is done iteratively until only one cluster that includes all items remains. Some hierarchical methods work in reverse, starting with all the items in a single cluster and then separating out items based on dissimilarity.

The *k*-means method uses a measure of *distance* between items to determine dissimilarity. The *Euclidean distance*, the squared difference between variable values of two items, is frequently used with the *k*-means method. Measures that focus on distance between clusters, the most significant variable difference, or minimizing the variance calculation within clusters can also be used.

For example, the *Manhattan distance*, also known as the *city block distance*, uses the absolute value of the difference between variable values of the items. This distance uses a calculation analogous to the distance an automobile would travel on roadways that form right angles, as many streets and avenues in Manhattan do. This distance would be most appropriate for clustering that uses medians, and not means, as the basis for forming clusters.

Because clustering methods calculate the measure of similarity or dissimilarity from variable values, clustering methods are inherently examples of unsupervised learning. However, semi-supervised techniques that use hints (see Section 13.2) are starting to be used on a wide basis.

WORKED-OUT PROBLEM 3   You wish to confirm an objective basis for the classification of iris plants that occur naturally in one part of Quebec. You collect measurements of the petal and sepal parts of 150 iris plants that botanists have previously identified as being from one of three varieties. The following table shows the data for the first five iris plants selected for this sample.

Iris Plants

Sepal Length	Sepal Width	Petal Length	Petal Width
5.1	3.5	1.4	0.2
4.9	3.0	1.4	0.2
4.7	3.2	1.3	0.2
4.6	3.1	1.5	0.2
5.0	3.6	1.4	0.2

The results of hierarchical clustering can be visualized as a *dendogram*, which is a type of tree diagram. The Figure 13.3 dendogram for the iris plant sample has many levels, each representing the combination of two clusters, or items. When a dendogram has many levels, a distance diagram, which Figure 13.3 also contains, helps determine at which level a dendogram should be used.

FIGURE 13.3

Dendogram and distance diagram for the iris plant sample

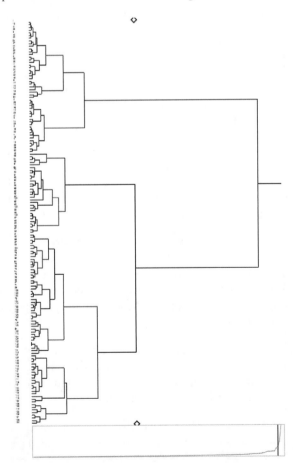

You use a distance diagram to find the level that causes the distance diagram plot line to take a steep turn (or *change its slope severely*, in geometric terms). For the Figure 13.3 iris plant distance diagram, this steep turn occurs at the second level from the right, when only three clusters remain.

The black vertical line on the distance diagram marks that level, and the small carets at the top and bottom of the dendrogram point to that level. Looking for the steep turn in the distance diagram is sometimes called the *elbow method* for the shape of an arm when bent at the elbow. After the "elbow," the distance diagram shows little change, suggesting that clustering into fewer than three clusters is not useful.

The k-means method can also be applied to the iris plant sample. With k set to 3, one of the three clusters has variable means very different from the other two (see Figure 13.4 top). That difference can be visualized in a *biplot*, a special type of scatter plot. In the Figure 13.4 biplot, the "very different" cluster is colored blue and appears apart from the other two. The other two clusters overlap in a 2D visualization, even though their means are distinct. This often occurs, and 3D visualizations (not shown) can be used to provide clearer visual distinctions among overlapping clusters.

FIGURE 13.4

Cluster means table and biplot for the iris plant sample

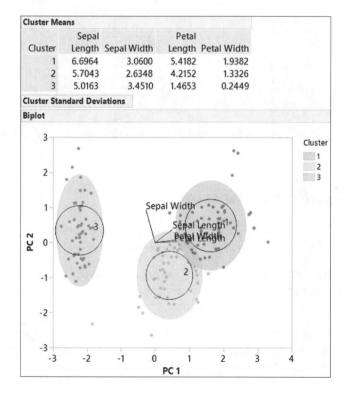

Cluster Means

Cluster	Sepal Length	Sepal Width	Petal Length	Petal Width
1	6.6964	3.0600	5.4182	1.9382
2	5.7043	2.6348	4.2152	1.3326
3	5.0163	3.4510	1.4653	0.2449

Cluster Standard Deviations

Biplot

MORE ABOUT THE IRIS DATA SET

The iris plant data plays an important role in the history of statistics and analytics. The statistician R. A. Fisher famously used this data set to demonstrate a number of analytical techniques that he devised. Because of that usage, the data is sometimes called *Fisher's iris data set*, even though the data were collected by another researcher!

Fisher's iris data set includes the varietal identity of each plant sampled. With this variable included, the data set illustrates the complementary nature of clustering and classification. With plant identification labels included, classification methods would be used to group the data. Without such labels, as in WORKED-OUT PROBLEM 3, clustering can be used to group the data.

software solution

Clustering

The clustering results that WORKED-OUT PROBLEM 3 use were created using the Hierarchical Cluster and K Means Cluster procedures of JMP, statistical software distributed by SAS Institute Inc.

How-Tos

Readers interested in exploring clustering using JMP can download a 30-day trial version of JMP at **www.jmp.com**. Also available at that website are tutorials and webinars about JMP and the full documentation for using the clustering procedures.

13.5 Association Analysis

CONCEPT   Methods that find sets of variable values that tend to co-occur or the rules that specify the co-occurrence of the values of two or more variables.

INTERPRETATION   Association methods use one of several techniques for finding co-occurrences. The models these methods create can be expressed as probabilistic if-then rules, such as "If item A occurs, then item B is likely to occur."

One way to establish these rules is to consider all possible rules for a set of items. That seems simple, but considering all possible rules can become time-consuming because for n number of items, the number of all possible rules is $3n - 2^{(n+1)} + 1$. For a trivial set of two items, there are only 2 rules, but for a group of 10 items, there are thousands of possible rules to consider (57,002, to be exact)!

Association methods use insights and shortcuts to reduce the number of possible rules to consider. The *Apriori method* looks to prune rules that concern associations of low frequency. For example, in a set of five items A, B, C, D, and E, if the set AB rarely occurs, then all sets that include AB—including ABC, ABCD, and ABCDE—will also be low frequency. This enables the *pruning* of rules that involve those sets, thereby reducing the number of rules that need to be considered.

The Apriori method and the *FP-Growth method*, which contains refinements suitable for certain data sources, are commonly used in *market basket analysis*, which is perhaps the most well-known application of association analysis.

Market Basket Analysis

CONCEPT   The application of association analysis to retail customer purchasing behavior.

INTERPRETATION   Market basket analysis takes its name from the baskets or carts that shoppers use to gather items for purchase and in which customers can purchase multiple items in one retail transaction. This type of analysis examines retail transactions based on the items purchased in each transaction. The analysis can discover which items are likely to be purchased together or which group of items is likely to be purchased with another group of items.

For retailers such as grocery stores or general discounters such as Walmart or Target, market basket analysis guides the placement of items in a store and provides actionable information about which sets of items should be placed on sale at the same time. For example, if a store discovers that product A and product B are typically bought together, the store can place one of the items on sale, knowing that a customer buying product A is likely to buy product B.

Examples of this type of analysis usually focus on tangible items such as the example in the preceding paragraph. However, market basket analysis can also be applied to intangible items, such as banking or financial services that might be "bought" together in one transaction.

WORKED-OUT PROBLEM 4   You have been asked to pilot a market basket analysis for a local convenience store that sells food and beverage items. Managers would like to know more about co-occurrences of categories of items to better plan future promotions. You collect the item details of 500 transactions from a typical sales day.

Convenience Store Analysis

The analysis creates a *frequent item set* table that lists the items frequently purchased together. In the Figure 13.5 JMP (partial) frequent item set table, the Support column contains the percentage of occurrence of each frequent item set in the data being analyzed. Among other things, the partial results reveal that purchasers of beer often also frequently buy energy drinks.

FIGURE 13.5

JMP frequent
item set table
(partial) for the
convenience
store market
basket
analysis

Frequent Item Sets	
Item Set ⌃	Support
{beer, crab cakes, ice cream}	15%
{beer, crab cakes}	18%
{beer, energy drink, ice cream, pasta, salty snacks}	13%
{beer, energy drink, ice cream, pasta}	13%
{beer, energy drink, ice cream, salty snacks}	17%
{beer, energy drink, ice cream}	19%
{beer, energy drink, pasta, salty snacks}	14%
{beer, energy drink, pasta}	14%
{beer, energy drink, salads, salty snacks, wine}	11%
{beer, energy drink, salads, salty snacks}	13%
{beer, energy drink, salads, wine}	11%
{beer, energy drink, salads}	13%
{beer, energy drink, salty snacks, wine}	12%
{beer, energy drink, salty snacks}	24%
{beer, energy drink, wine}	13%
{beer, energy drink}	26%
{beer, fish, ice cream, pasta}	12%
{beer, fish, ice cream, pizza}	12%

Managers are somewhat surprised by these co-occurrences because they had
assumed that beer and energy drinks were purchased by different types of
customers. Combinations that include beer and energy drinks define a number
of the association rules that the analysis defines in the Figure 13.6 (partial) table
of results.

FIGURE 13.6

JMP
association
rules table
(partial) for the
convenience
store market
basket
analysis

Rules			
Rule			
Condition	⌃ Consequent	Confidence	Lift
beer, crab cakes, ice cream	bottled water, chicken	79%	5.00
beer, energy drink	salty snacks	92%	1.84
beer, energy drink	ice cream, salty snacks	64%	2.27
beer, energy drink	pasta	56%	1.37
beer, energy drink	pasta, salty snacks	53%	2.72
beer, energy drink	ice cream, pasta	52%	1.69
beer, energy drink	salads, salty snacks	50%	2.60
beer, energy drink	ice cream, pasta, salty snacks	49%	3.30
beer, energy drink	salty snacks, wine	48%	1.99
beer, energy drink	salads, wine	43%	1.71
beer, energy drink	salads, salty snacks, wine	43%	3.19
beer, energy drink, ice cream	salty snacks	89%	1.78
beer, energy drink, ice cream	pasta	72%	1.77
beer, energy drink, ice cream	pasta, salty snacks	68%	3.49
beer, energy drink, pasta	salty snacks	94%	1.89
beer, energy drink, pasta	ice cream	93%	1.27
beer, energy drink, pasta	ice cream, salty snacks	88%	3.08
beer, energy drink, salads	salty snacks	98%	1.97
beer, energy drink, salads	wine	85%	2.05
beer, energy drink, salads	salty snacks, wine	85%	3.51
beer, energy drink, salty snacks	pasta	57%	1.40
beer, energy drink, salty snacks	ice cream, pasta	53%	1.72
beer, energy drink, salty snacks	wine	52%	1.26
beer, energy drink, salty snacks	salads, wine	47%	1.85
beer, energy drink, wine	salty snacks	98%	1.97
beer, energy drink, wine	salads	89%	1.88
beer, energy drink, wine	salads, salty snacks	89%	4.58
beer, fish	pizza	74%	2.14

In the results table, the Confidence column shows the proportion of transactions that meet the listed condition and contain the "consequent" categories. The Lift column reflects how much more (or less) likely than random chance is the occurrence. Values greater than 1 are more likely to be due to associations. From the results, managers can conclude that promotions involving both beer and energy drinks probably should not be promoted together as the co-occurrence between the two is common. Because the salty snacks category is a consequent of a number of rules involving beer and energy drinks, items in that category could be offered at full price when beer or energy drink promotions occur without incurring a loss of sales volume in the salty snacks category.

FURTHER INTERPRETATION   The Figure 13.6 confidence and lift values need further interpretation—as results of predictive analysis often do. For association analysis, rules with a confidence score over 90% (or even 100%) may be caused by the low frequency of the association. For the managers of the convenience store, a rule with a 100% confidence score that applies to only 5% of all transactions may not be as important as a rule with an 83% score that applies to 55% of all transactions. Likewise, selecting rules solely based on lift scores would not automatically be the best choice.

software solution

Association Analysis

The association analysis results that WORKED-OUT PROBLEM 4 use were created using the Association Analysis procedure of JMP, statistical software distributed by SAS Institute Inc.

How-Tos

Readers interested in exploring clustering using JMP can download a 30-day trial version of JMP at **www.jmp.com**. Also available at that website are tutorials and webinars about JMP and the full documentation for using the clustering procedures.

One-Minute Summary

Predictive analytics

- Use a target-base method to predict the value of one variable based on the value of other variables.

- Use a classification method to group items into two or more groups or classes.
  - With a categorical dependent variable, use a classification tree.
  - With a numerical dependent variable, use a regression tree.
- Use a clustering method to discover grouping of similar items for data that contains no dependent variable (no labeled items).
- Use an association analysis to find sets of variable values that tend to co-occur or the rules that specify the co-occurrence of the values of two or more variables.

Test Yourself

1. A financial services company wants to be able to predict whether a cardholder would be willing to upgrade to a premium credit card. Among the independent variables to be considered are the monthly amount charged on the current credit card and the number of credit cards owned. Which predictive analytics method would be best to use for this problem?
   a.   classification trees
   b.   regression trees
   c.   clustering method
   d.   association analysis

2. A winery wants to study the factors involved in the quality rating of a wine. Data have been collected on various characteristics that can influence the quality rating of the wine. Which predictive analytics method would be best to use for this problem?
   a.   classification trees
   b.   regression trees
   b.   clustering method
   d.   association analysis

3. A historical researcher wants to better understand the demographic factors that might have increased the likelihood of surviving the sinking of the RMS *Titanic* in 1912. For each passenger on the ship, data on the age, gender, class of service booked, and whether the passenger was a survivor are collected. Which of the following analytics methods should you use to develop a model to predict whether a passenger survived?
   a.   classification trees
   b.   regression trees
   c.   clustering method
   d.   association analysis

4. A basketball analytics specialist wants to develop a model to predict the number of wins achieved by an NBA team. Among the independent variables used are field goal percentage, opponents' field goal percentage, three-point field goal percentage, opponents' three-point field goal percentage, rebounds, and turnovers. Which predictive analytics method would be best to use for this problem?

    a.   classification trees

    b.   regression trees

    c.   clustering method

    d.   association analysis

5. A restaurant owner wants to study the perception that customers have of various main course choices. The owner asks customers to rate each main course by various characteristics, including spiciness, taste, healthfulness, and calories. The owner wants to determine which main courses have similar characteristics. Which predictive analytics method would be best to use for this problem?

    a.   target-based method

    b.   classification method

    c.   clustering method

    d.   association analysis

6. A restaurant owner wants to better understand which combinations of appetizers, main courses, desserts, and beverages are most likely to be ordered together during weekend dinner hours. Which predictive analytics method would be best to use for this problem?

    a.   target-based method

    b.   classification method

    c.   clustering method

    d.   association analysis

7. A baseball analytics specialist wants to study the similarities and differences among the teams in a recent season. The specialist obtains offensive and pitching statistics for each team, including runs scored per game, home runs, batting average, runs allowed, earned run average, saves, and opponents' batting average. Which predictive analytics method would be best to use for this problem?

    a.   target-based method

    b.   classification method

    c.   clustering

    d.   association analysis

8. A baseball analytics specialist wants to develop a model that predicts the number of team wins based on various offensive and pitching statistics for the team. Which predictive analytics method would be best to use for this problem?

    a.   target-based method
    b.   classification method
    c.   clustering method
    d.   association analysis

Answer True or False:

9. All predictive analytics methods make it possible to discover patterns in the data being analyzed.

10. Tree induction methods are examples of unsupervised learning.

11. Clustering methods are examples of unsupervised learning.

12. Pruning is done in a clustering method to improve the clusters.

13. In leave-one-out cross-validation, one instance or item of data is never analyzed.

14. In cluster analysis, items in a group are more like other items in their group than they are like items in another group.

15. Market basket analysis is a well-known application of tree induction.

16. Methods that use unsupervised learning may create models that cannot be easily explained or mapped to real-world knowledge.

17. Models created using predictive analytics methods assist but do not replace human decision making.

18. Regression trees are an example of classification models.

19. The Apriori method is an example of a classification method.

20. Association rules with the highest confidence, or lift values, may not be the most useful rules to a decision maker.

Answers to Test Yourself

1. a	11. True
2. b	12. False
3. a	13. False
4. b	14. True
5. c	15. False
6. d	16. True
7. a	17. True
8. d	18. False
9. True	19. False
10. False	20. True

References

1. Berenson, M. L., D. M. Levine, K. A. Szabat, and D. Stephan. *Basic Business Statistics: Concepts and Applications*, 15th edition. Hoboken, NJ: Pearson Education, 2023.

2. Han, J., J. Pei, and M. Kamber. *Data Mining: Concepts and Techniques*, 3rd edition. Burlington, MA: Morgan Kaufmann, 2011.

3. KDNuggets. "Market Basket Analysis: A Tutorial," https://bit.ly/3DZqREl.

4. Kaushik, S. "An Introduction to Clustering and Different Methods of Clustering," Analytics Vidhya blog, https://bit.ly/3p0uV0G.

5. Klimberg, R., and B. D. McCullough. *Fundamentals of Predictive Analytics with JMP*, 2nd edition. Cary, NC: SAS Institute, 2017.

6. Provost F., and T. Fawcett. *Data Science for Business*. Sebastopol, CA: O'Reilly Media, 2013.

APPENDIX

A

Microsoft Excel Operation and Configuration

This appendix assists you in operating and configuring your copy of Microsoft Excel for use with this book, including the conventions other appendixes use to describe keystroke and mouse user operations.

a.1 Conventions for Keystroke and Mouse Operations

The spreadsheet operation instructions in this book use a standard vocabulary to describe keystroke and mouse (pointer) operations. Keys are always named by their legends. For example, the instruction "press Enter" means to press the key with the legend **Enter**.

For mouse operation, this book uses **click** and **select** and, less frequently, **check**, **right-click**, and **double-click**. Click means to move the pointer over an object and press the primary mouse button. Select means to either find and highlight a named choice from a pull-down list or fill in an option button (also known as a radio button) associated with that choice. Check means to fill in the check box of a choice by clicking in its empty check box. Right-click means to press the secondary mouse button (or to hold down the Control key and press the mouse button, if using a one-button mouse). Double-click means to press the primary mouse button rapidly twice to select an object directly.

a2 **Microsoft Excel Technical Configuration**

The instructions in this book for using Microsoft Excel assume no special technical settings. If you plan to use any of the Analysis ToolPak Tips in Appendix E, you need to make sure that the Analysis ToolPak add-in has been installed in your copy of Microsoft Excel.

To check for the presence of the Analysis ToolPak add-in, if you use Microsoft Excel with Microsoft Windows, follow these steps:

1. Select **File → Options**.

In the Excel Options dialog box:

2. Click **Add-Ins** in the left pane and look for the entry **Analysis ToolPak** in the right pane, under **Active Application Add-ins**.

3. If the entry appears, click **OK**.

4. If the entry does not appear in the Active Application Add-Ins list, select **Excel Add-Ins** from the **Manage** drop-down list and then click **Go**.

5. In the Add-Ins dialog box, check **Analysis ToolPak** in the **Add-ins available** list and click **OK**. If Analysis ToolPak does not appear in the add-ins available list, rerun the Microsoft Office setup program to add the Analysis ToolPak add-in.

6. Close and restart Excel.

To check for the presence of the Analysis ToolPak add-in, if you use Microsoft Excel for Mac, follow these steps:

1. Select **Tools → Excel Add-ins**.

2. In the **Add-ins available** list, check the **Analysis ToolPak** check box.

3. If the entry appears, click **OK**.

4. If the entry does not appear, click **Browse** to locate the add-in. if the Analysis ToolPak cannot be found, click **Yes** when asked to install the add-in.

5. Close and restart Excel.

Review of Arithmetic and Algebra

The authors understand that wide differences exist in the mathematical backgrounds of readers of this book. Some readers might have taken various courses in algebra, calculus, and matrix algebra, while others might never have taken any mathematics courses or took such courses a long time ago.

Because this book emphasizes statistical concepts and the interpretation of spreadsheet results, no mathematical prerequisite other than elementary arithmetic and algebra is needed. Use the results of the assessment quiz to help you identify the arithmetic and algebraic concepts of this appendix that you may need to study and review.

Assessment Quiz

Part 1

Fill in the correct answer.

1. $\dfrac{\frac{1}{2}}{3} =$

2. $(0.4)^2 =$

3. $1 + \dfrac{2}{3} =$

4. $\left(\dfrac{1}{3}\right)^{(4)} =$

5. $\dfrac{1}{5} =$ (in decimals)

6. $1 - (-0.3) =$

7. $4 \times 0.2 \times (-8) =$

8. $\left(\dfrac{1}{4} \times \dfrac{2}{3}\right) =$

9. $\left(\dfrac{1}{100}\right) + \left(\dfrac{1}{200}\right) =$

10. $\sqrt{16} =$

Part 2

Select the correct answer.

1. If $a = bc$, then $c =$

    a. ab

    b. b/a

    c. a/b

    d.   none of the above

2. If $x + y = z$, then $y =$

    a. z/x

    b. $z + x$

    c. $z - x$

    d.   none of the above

3. $(x^3)(x^2) =$

    a. x^5

    b. x^6

    c. x^1

    d.   none of the above

4. $x^0 =$

    a. x

    b.   1

    c.   0

    d.   none of the above

5. $x(y - z) =$

    a. $xy - xz$

    b. $xy - z$

    c. $(y - z)/x$

    d.   none of the above

6. $(x + y)/z =$

    a. $(x/z) + y$

    b. $(x/z) + (y/z)$

    c. $x + (y/z)$

    d.   none of the above

7. $x /(y + z) =$

    a. $(x/y) + (1/z)$

    b. $(x/y) + (x/z)$

    c. $(y +z)/ x$

    d.   none of the above

8. If $x = 10$, $y = 5$, $z = 2$, and $w = 20$, then $(xy - z^2)/w =$

    a.   5

    b.   2.3

    c.   46

    d.   none of the above

9. $(8x^4)/(4x^2) =$

    a. $2x^2$

    b.   2

    c. $2x$

    d.   none of the above

10. $\sqrt{\dfrac{X}{Y}} =$

    a. $\sqrt{Y} / \sqrt{X}$

    b. $\sqrt{1} / \sqrt{XY}$

    c. $\sqrt{X} / \sqrt{Y}$

    d.   none of the above

The answers to both parts of the quiz appear at the end of this appendix.

Symbols

Each of the four basic arithmetic operations—addition, subtraction, multiplication, and division—is indicated by a symbol.

$+$ add

$\times$ or $\cdot$ multiply

$-$ subtract

$\div$ or $/$ divide

In addition to these operations, the following symbols are used to indicate equality or inequality:

$=$ equals

$\neq$ not equal

$\cong$ approximately equal to

$>$ greater than

$<$ less than

$\geq$ greater than or equal to

$\leq$ less than or equal to

Addition

Addition refers to the summation or accumulation of a set of numbers. In adding numbers, the two basic laws are the commutative law and the associative law.

The **commutative law** of addition states that the order in which numbers are added is irrelevant. This can be seen in the following two examples:

$$1 + 2 = 3 \qquad 2 + 1 = 3$$
$$x + y = z \qquad y + x = z$$

In each example, the number that was listed first and the number that was listed second did not matter.

The **associative law** of addition states that in adding several numbers, any subgrouping of the numbers can be added first, last, or in the middle. You can see this in the following examples:

$$2 + 3 + 6 + 7 + 4 + 1 = 23$$
$$(5) + (6 + 7) + 4 + 1 = 23$$
$$5 + 13 + 5 = 23$$
$$5 + 6 + 7 + 4 + 1 = 23$$

In each of these examples, the order in which the numbers have been added has no effect on the results.

Subtraction

The process of subtraction is the opposite or inverse of addition. The operation of subtracting 1 from 2 (that is, $2 - 1$) means that one unit is to be taken away from two units, leaving a remainder of one unit. In contrast to addition, the commutative and associative laws do not hold for subtraction. Therefore, as indicated in the following examples,

$$8 - 4 = 4 \quad\quad \text{but} \quad\quad 4 - 8 = -4$$
$$3 - 6 = -3 \quad\quad \text{but} \quad\quad 6 - 3 = 3$$
$$8 - 3 - 2 = 3 \quad\quad \text{but} \quad\quad 3 - 2 - 8 = -7$$
$$9 - 4 - 2 = 3 \quad\quad \text{but} \quad\quad 2 - 4 - 9 = -11$$

When subtracting negative numbers, remember that the same result occurs when subtracting a negative number as when adding a positive number. Thus,

$$4 - (-3) = +7 \quad\quad 4 + 3 = 7$$
$$8 - (-10) = +18 \quad\quad 8 + 10 = 18$$

Multiplication

The operation of multiplication is a shortcut method of addition when the same number is to be added several times. For example, if 7 is added three times (7 + 7 + 7), you could multiply 7 times 3 to get the product of 21.

In multiplication as in addition, the commutative laws and associative laws are in operation so that:

$$a \times b = b \times a$$
$$4 \times 5 = 5 \times 4 = 20$$
$$(2 \times 5) \times 6 = 10 \times 6 = 60$$

A third law of multiplication, the **distributive law**, applies to the multiplication of one number by the sum of several numbers. Here,

$$a(b + c) = ab + ac$$
$$2(3 + 4) = 2(7) = 2(3) + 2(4) = 14$$

The resulting product is the same regardless of whether b and c are summed and multiplied by a or a is multiplied by b and by c and the two products are added together.

You also need to remember that when multiplying negative numbers, a negative number multiplied by a negative number equals a positive number. Thus,

$$(-a) \times (-b) = ab$$
$$(-5) \times (-4) = +20$$

Division

Just as subtraction is the opposite of addition, division is the opposite or inverse of multiplication. Division can be viewed as a shortcut to subtraction. When you divide 20 by 4, you are actually determining the number of times that 4 can be subtracted from 20. In general, however, the number of times one number can be divided by another may not be an exact integer value because there could be a remainder. For example, if you divide 21 by 4, the answer is 5 with a remainder of 1, or $5\frac{1}{4}$.

As in the case of subtraction, neither the commutative nor associative law of addition and multiplication holds for division:

$$a \div b \neq b \div a$$
$$9 \div 3 \neq 3 \div 9$$
$$6 \div (3 \div 2) = 4$$
$$(6 \div 3) \div 2 = 1$$

The distributive law holds only when the numbers to be added are contained in the numerator, not the denominator. Thus,

$$\frac{a+b}{c} = \frac{a}{c} + \frac{b}{c} \quad \text{but} \quad \frac{a}{b+c} \neq \frac{a}{b} + \frac{a}{c}$$

For example,

$$\frac{1}{2+3} = \frac{1}{5} \quad \text{but} \quad \frac{1}{2+3} \neq \frac{1}{2} + \frac{1}{3}$$

The last important property of division states that if the numerator and the denominator are multiplied or divided by the same number, the resulting quotient is not affected. Therefore,

$$\frac{80}{40} = 2$$

then

$$\frac{5(80)}{5(40)} = \frac{400}{200} = 2$$

and

$$\frac{80 \div 5}{40 \div 5} = \frac{16}{8} = 2$$

Fractions

A fraction is a number that consists of a combination of whole numbers and/or parts of whole numbers. For instance, the fraction 1/3 consists of only one portion of a number, while the fraction 7/6 consists of the whole number 1 plus the fraction 1/6. Each of the operations of addition, subtraction, multiplication, and division can be used with fractions. When adding or subtracting fractions, you must find the lowest common denominator for each fraction prior to adding or subtracting them. Thus, in adding 1/3 + 1/5, the lowest common denominator is 15, so

$$\frac{5}{15} + \frac{3}{15} = \frac{8}{15}$$

In subtracting 1/4 − 1/6, the same principles applies, so that the lowest common denominator is 12, producing a result of

$$\frac{3}{12} - \frac{2}{12} = \frac{1}{12}$$

Multiplying and dividing fractions does not have the lowest common denominator requirement associated with adding and subtracting fractions. Thus, if a/b is multiplied by c/d, the result is ac/bd.

The resulting numerator, ac, is the product of the numerators a and c, while the denominator, bd, is the product of the two denominators b and d. The resulting fraction can sometimes be reduced to a lower term by dividing the numerator and denominator by a common factor. For example, taking

$$\frac{2}{3} \times \frac{6}{7} = \frac{12}{21}$$

and dividing the numerator and denominator by 3 produces the result 4/7.

Division of fractions can be thought of as the inverse of multiplication, so the divisor can be inverted and multiplied by the original fraction. Thus,

$$\frac{9}{5} \div \frac{1}{4} = \frac{9}{5} \times \frac{4}{1} = \frac{36}{5}$$

The division of a fraction can also be thought of as a way of converting the fraction to a decimal number. For example, the fraction 2/5 can be converted to a decimal number by dividing its numerator, 2, by its denominator, 5, to produce the decimal number 0.40.

Exponents and Square Roots

Exponentiation (raising a number to a power) provides a shortcut in writing numerous multiplications. For example, $2 \times 2 \times 2 \times 2 \times 2$ can be written as $2^5 = 32$. The 5 represents the exponent (or power) of the number 2, telling you that 2 is to be multiplied by itself five times.

Several rules can be used for multiplying or dividing numbers that contain exponents.

Rule 1: $x^a \cdot x^b = x^{(a+b)}$

If two numbers involving a power of the same number are multiplied, the product is the same number raised to the sum of the powers.

$$4^2 \cdot 4^3 = (4 \cdot 4)(4 \cdot 4 \cdot 4) = 4^5$$

Rule 2: $(x^a)^b = x^{ab}$

If you take the power of a number that is already taken to a power, the result is a number that is raised to the product of the two powers. For example,

$$(4^2)^3 = (4^2)(4^2)(4^2) = 4^6$$

Rule 3: $\dfrac{X^a}{X^b} = X^{(a-b)}$

If a number raised to a power is divided by the same number raised to a power, the quotient is the number raised to the difference of the powers. Thus,

$$\frac{3^5}{3^3} = \frac{3 \cdot 3 \cdot 3 \cdot 3 \cdot 3}{3 \cdot 3 \cdot 3} = 3^2$$

If the denominator has a higher power than the numerator, the resulting quotient is a negative power. Thus,

$$\frac{3^3}{3^5} = \frac{3 \cdot 3 \cdot 3}{3 \cdot 3 \cdot 3 \cdot 3 \cdot 3} = \frac{1}{3^2} = 3^{-2} = \frac{1}{9}$$

If the difference between the powers of the numerator and denominator is 1, the result is the number itself. In other words, $x^1 = x$. For example,

$$\frac{3^3}{3^2} = \frac{3 \cdot 3 \cdot 3}{3 \cdot 3} = 3^{-1} = 3$$

If, however, no difference exists in the power of the numbers in the numerator and denominator, the result is 1. Thus,

$$\frac{x^a}{x^a} = x^{a-a} = x^0 = 1$$

Therefore, any number raised to the zero power equals 1. For example,

$$\frac{3^3}{3^3} = \frac{3 \cdot 3 \cdot 3}{3 \cdot 3 \cdot 3} = 3^0 = 1$$

The square root represented by the symbol $\sqrt{}$ is a special power of a number, the ½ power. It indicates the value that when multiplied by itself, will produce the original number.

Equations

In statistics, many formulas are expressed as equations where one unknown value is a function of another value. Thus, it is important that you know how to manipulate equations into various forms. The rules of addition, subtraction, multiplication, and division can be used to work with equations. For example, the equation

$$x - 2 = 5$$

can be solved for x by adding 2 to each side of the equation. This results in

$x - 2 + 2 = 5 + 2$. Therefore, $x = 7$.

If

$$x + y = z$$

you could solve for x by subtracting y from both sides of the equation so that

$x + y - y = z - y$. Therefore, $x = z - y$.

If the product of two variables is equal to a third variable, such as

$$x y = z,$$

you can solve for x by dividing both sides of the equation by y. Thus,

$$\frac{xy}{y} = \frac{z}{y}$$

$$x = \frac{z}{y}$$

Conversely, if

$$\frac{x}{y} = z$$

you can solve for x by multiplying both sides of the equation by y.

$$\frac{xy}{y} = zy$$
$$x = 2y$$

To summarize, the various operations of addition, subtraction, multiplication, and division can be applied to equations as long as the same operation is performed on each side of the equation, thereby maintaining the equality.

Answers to Quiz

Part 1

1. $3/2$
2. 0.16
3. $5/3$
4. $1/81$
5. 0.20
6. 1.30
7. −6.4
8. $+1/6$
9. $3/200$
10. 4

Part 2

1. c
2. c
3. a
4. b
5. a
6. b
7. d
8. b
9. a
10. c

APPENDIX

C

Statistical Tables

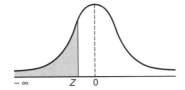

TABLE C.1   The Cumulative Standardized Normal Distribution

Entry represents area under the cumulative standardized normal distribution from $-\infty$ to Z.

Z	0.00	0.01	0.02	0.03	0.04	0.05	0.06	0.07	0.08	0.09
−3.9	0.00005	0.00005	0.00004	0.00004	0.00004	0.00004	0.00004	0.00004	0.00003	0.00003
−3.8	0.00007	0.00007	0.00007	0.00006	0.00006	0.00006	0.00006	0.00005	0.00005	0.00005
−3.7	0.00011	0.00010	0.00010	0.00010	0.00009	0.00009	0.00008	0.00008	0.00008	0.00008
−3.6	0.00016	0.00015	0.00015	0.00014	0.00014	0.00013	0.00013	0.00012	0.00012	0.00011
−3.5	0.00023	0.00022	0.00022	0.00021	0.00020	0.00019	0.00019	0.00018	0.00017	0.00017
−3.4	0.00034	0.00032	0.00031	0.00030	0.00029	0.00028	0.00027	0.00026	0.00025	0.00024
−3.3	0.00048	0.00047	0.00045	0.00043	0.00042	0.00040	0.00039	0.00038	0.00036	0.00035
−3.2	0.00069	0.00066	0.00064	0.00062	0.00060	0.00058	0.00056	0.00054	0.00052	0.00050
−3.1	0.00097	0.00094	0.00090	0.00087	0.00084	0.00082	0.00079	0.00076	0.00074	0.00071
−3.0	0.00135	0.00131	0.00126	0.00122	0.00118	0.00114	0.00111	0.00107	0.00103	0.00100
−2.9	0.0019	0.0018	0.0018	0.0017	0.0016	0.0016	0.0015	0.0015	0.0014	0.0014
−2.8	0.0026	0.0025	0.0024	0.0023	0.0023	0.0022	0.0021	0.0021	0.0020	0.0019
−2.7	0.0035	0.0034	0.0033	0.0032	0.0031	0.0030	0.0029	0.0028	0.0027	0.0026
−2.6	0.0047	0.0045	0.0044	0.0043	0.0041	0.0040	0.0039	0.0038	0.0037	0.0036
−2.5	0.0062	0.0060	0.0059	0.0057	0.0055	0.0054	0.0052	0.0051	0.0049	0.0048
−2.4	0.0082	0.0080	0.0078	0.0075	0.0073	0.0071	0.0069	0.0068	0.0066	0.0064
−2.3	0.0107	0.0104	0.0102	0.0099	0.0096	0.0094	0.0091	0.0089	0.0087	0.0084
−2.2	0.0139	0.0136	0.0132	0.0129	0.0125	0.0122	0.0119	0.0116	0.0113	0.0110
−2.1	0.0179	0.0174	0.0170	0.0166	0.0162	0.0158	0.0154	0.0150	0.0146	0.0143
−2.0	0.0228	0.0222	0.0217	0.0212	0.0207	0.0202	0.0197	0.0192	0.0188	0.0183

TABLE C.1   313

Z	0.00	0.01	0.02	0.03	0.04	0.05	0.06	0.07	0.08	0.09
−1.9	0.0287	0.0281	0.0274	0.0268	0.0262	0.0256	0.0250	0.0244	0.0239	0.0233
−1.8	0.0359	0.0351	0.0344	0.0336	0.0329	0.0322	0.0314	0.0307	0.0301	0.0294
−1.7	0.0446	0.0436	0.0427	0.0418	0.0409	0.0401	0.0392	0.0384	0.0375	0.0367
−1.6	0.0548	0.0537	0.0526	0.0516	0.0505	0.0495	0.0485	0.0475	0.0465	0.0455
−1.5	0.0668	0.0655	0.0643	0.0630	0.0618	0.0606	0.0594	0.0582	0.0571	0.0559
−1.4	0.0808	0.0793	0.0778	0.0764	0.0749	0.0735	0.0721	0.0708	0.0694	0.0681
−1.3	0.0968	0.0951	0.0934	0.0918	0.0901	0.0885	0.0869	0.0853	0.0838	0.0823
−1.2	0.1151	0.1131	0.1112	0.1093	0.1075	0.1056	0.1038	0.1020	0.1003	0.0985
−1.1	0.1357	0.1335	0.1314	0.1292	0.1271	0.1251	0.1230	0.1210	0.1190	0.1170
−1.0	0.1587	0.1562	0.1539	0.1515	0.1492	0.1469	0.1446	0.1423	0.1401	0.1379
−0.9	0.1841	0.1814	0.1788	0.1762	0.1736	0.1711	0.1685	0.1660	0.1635	0.1611
−0.8	0.2119	0.2090	0.2061	0.2033	0.2005	0.1977	0.1949	0.1922	0.1894	0.1867
−0.7	0.2420	0.2388	0.2358	0.2327	0.2296	0.2266	0.2236	0.2206	0.2177	0.2148
−0.6	0.2743	0.2709	0.2676	0.2643	0.2611	0.2578	0.2546	0.2514	0.2482	0.2451
−0.5	0.3085	0.3050	0.3015	0.2981	0.2946	0.2912	0.2877	0.2843	0.2810	0.2776
−0.4	0.3446	0.3409	0.3372	0.3336	0.3300	0.3264	0.3228	0.3192	0.3156	0.3121
−0.3	0.3821	0.3783	0.3745	0.3707	0.3669	0.3632	0.3594	0.3557	0.3520	0.3483
−0.2	0.4207	0.4168	0.4129	0.4090	0.4052	0.4013	0.3974	0.3936	0.3897	0.3859
−0.1	0.4602	0.4562	0.4522	0.4483	0.4443	0.4404	0.4364	0.4325	0.4286	0.4247
−0.0	0.5000	0.4960	0.4920	0.4880	0.4840	0.4801	0.4761	0.4721	0.4681	0.4641

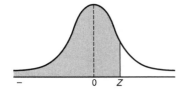

TABLE C.1   Continued

Entry represents area under the standardized normal distribution from $-\infty$ to Z.

Z	0.00	0.01	0.02	0.03	0.04	0.05	0.06	0.07	0.08	0.09
0.0	0.5000	0.5040	0.5080	0.5120	0.5160	0.5199	0.5239	0.5279	0.5319	0.5359
0.1	0.5398	0.5438	0.5478	0.5517	0.5557	0.5596	0.5636	0.5675	0.5714	0.5753
0.2	0.5793	0.5832	0.5871	0.5910	0.5948	0.5987	0.6026	0.6064	0.6103	0.6141
0.3	0.6179	0.6217	0.6255	0.6293	0.6331	0.6368	0.6406	0.6443	0.6480	0.6517
0.4	0.6554	0.6591	0.6628	0.6664	0.6700	0.6736	0.6772	0.6808	0.6844	0.6879
0.5	0.6915	0.6950	0.6985	0.7019	0.7054	0.7088	0.7123	0.7157	0.7190	0.7224
0.6	0.7257	0.7291	0.7324	0.7357	0.7389	0.7422	0.7454	0.7486	0.7518	0.7549
0.7	0.7580	0.7612	0.7642	0.7673	0.7704	0.7734	0.7764	0.7794	0.7823	0.7852
0.8	0.7881	0.7910	0.7939	0.7967	0.7995	0.8023	0.8051	0.8078	0.8106	0.8133
0.9	0.8159	0.8186	0.8212	0.8238	0.8264	0.8289	0.8315	0.8340	0.8365	0.8389
1.0	0.8413	0.8438	0.8461	0.8485	0.8508	0.8531	0.8554	0.8577	0.8599	0.8621
1.1	0.8643	0.8665	0.8686	0.8708	0.8729	0.8749	0.8770	0.8790	0.8810	0.8830
1.2	0.8849	0.8869	0.8888	0.8907	0.8925	0.8944	0.8962	0.8980	0.8997	0.9015
1.3	0.9032	0.9049	0.9066	0.9082	0.9099	0.9115	0.9131	0.9147	0.9162	0.9177
1.4	0.9192	0.9207	0.9222	0.9236	0.9251	0.9265	0.9279	0.9292	0.9306	0.9319
1.5	0.9332	0.9345	0.9357	0.9370	0.9382	0.9394	0.9406	0.9418	0.9429	0.9441
1.6	0.9452	0.9463	0.9474	0.9484	0.9495	0.9505	0.9515	0.9525	0.9535	0.9545
1.7	0.9554	0.9564	0.9573	0.9582	0.9591	0.9599	0.9608	0.9616	0.9625	0.9633
1.8	0.9641	0.9649	0.9656	0.9664	0.9671	0.9678	0.9686	0.9693	0.9699	0.9706
1.9	0.9713	0.9719	0.9726	0.9732	0.9738	0.9744	0.9750	0.9756	0.9761	0.9767
2.0	0.9772	0.9778	0.9783	0.9788	0.9793	0.9798	0.9803	0.9808	0.9812	0.9817
2.1	0.9821	0.9826	0.9830	0.9834	0.9838	0.9842	0.9846	0.9850	0.9854	0.9857
2.2	0.9861	0.9864	0.9868	0.9871	0.9875	0.9878	0.9881	0.9884	0.9887	0.9890

TABLE C.1   315

Z	0.00	0.01	0.02	0.03	0.04	0.05	0.06	0.07	0.08	0.09
2.3	0.9893	0.9896	0.9898	0.9901	0.9904	0.9906	0.9909	0.9911	0.9913	0.9916
2.4	0.9918	0.9920	0.9922	0.9925	0.9927	0.9929	0.9931	0.9932	0.9934	0.9936
2.5	0.9938	0.9940	0.9941	0.9943	0.9945	0.9946	0.9948	0.9949	0.9951	0.9952
2.6	0.9953	0.9955	0.9956	0.9957	0.9959	0.9960	0.9961	0.9962	0.9963	0.9964
2.7	0.9965	0.9966	0.9967	0.9968	0.9969	0.9970	0.9971	0.9972	0.9973	0.9974
2.8	0.9974	0.9975	0.9976	0.9977	0.9977	0.9978	0.9979	0.9979	0.9980	0.9981
2.9	0.9981	0.9982	0.9982	0.9983	0.9984	0.9984	0.9985	0.9985	0.9986	0.9986
3.0	0.99865	0.99869	0.99874	0.99878	0.99882	0.99886	0.99889	0.99893	0.99897	0.99900
3.1	0.99903	0.99906	0.99910	0.99913	0.99916	0.99918	0.99921	0.99924	0.99926	0.99929
3.2	0.99931	0.99934	0.99936	0.99938	0.99940	0.99942	0.99944	0.99946	0.99948	0.99950
3.3	0.99952	0.99953	0.99955	0.99957	0.99958	0.99960	0.99961	0.99962	0.99964	0.99965
3.4	0.99966	0.99968	0.99969	0.99970	0.99971	0.99972	0.99973	0.99974	0.99975	0.99976
3.5	0.99977	0.99978	0.99978	0.99979	0.99980	0.99981	0.99981	0.99982	0.99983	0.99983
3.6	0.99984	0.99985	0.99985	0.99986	0.99986	0.99987	0.99987	0.99988	0.99988	0.99989
3.7	0.99989	0.99990	0.99990	0.99990	0.99991	0.99991	0.99992	0.99992	0.99992	0.99992
3.8	0.99993	0.99993	0.99993	0.99994	0.99994	0.99994	0.99994	0.99995	0.99995	0.99995
3.9	0.99995	0.99995	0.99996	0.99996	0.99996	0.99996	0.99996	0.99996	0.99997	0.99997
4.0	0.99996832									
4.5	0.99999660									
5.0	0.99999971									
5.5	0.99999998									
6.0	0.99999999									

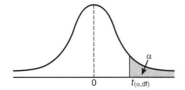

TABLE C.2   Critical Values of *t*

Degrees of Freedom	Upper-Tail Areas					
	0.25	0.10	0.05	0.025	0.01	0.005
1	1.0000	3.0777	6.3138	12.7062	31.8207	63.6574
2	0.8165	1.8856	2.9200	4.3027	6.9646	9.9248
3	0.7649	1.6377	2.3534	3.1824	4.5407	5.8409
4	0.7407	1.5332	2.1318	2.7764	3.7469	4.6041
5	0.7267	1.4759	2.0150	2.5706	3.3649	4.0322
6	0.7176	1.4398	1.9432	2.4469	3.1427	3.7074
7	0.7111	1.4149	1.8946	2.3646	2.9980	3.4995
8	0.7064	1.3968	1.8595	2.3060	2.8965	3.3554
9	0.7027	1.3830	1.8331	2.2622	2.8214	3.2498
10	0.6998	1.3722	1.8125	2.2281	2.7638	3.1693
11	0.6974	1.3634	1.7959	2.2010	2.7181	3.1058
12	0.6955	1.3562	1.7823	2.1788	2.6810	3.0545
13	0.6938	1.3502	1.7709	2.1604	2.6503	3.0123
14	0.6924	1.3450	1.7613	2.1448	2.6245	2.9768
15	0.6912	1.3406	1.7531	2.1315	2.6025	2.9467
16	0.6901	1.3368	1.7459	2.1199	2.5835	2.9208
17	0.6892	1.3334	1.7396	2.1098	2.5669	2.8982
18	0.6884	1.3304	1.7341	2.1009	2.5524	2.8784
19	0.6876	1.3277	1.7291	2.0930	2.5395	2.8609
20	0.6870	1.3253	1.7247	2.0860	2.5280	2.8453
21	0.6864	1.3232	1.7207	2.0796	2.5177	2.8314
22	0.6858	1.3212	1.7171	2.0739	2.5083	2.8188
23	0.6853	1.3195	1.7139	2.0687	2.4999	2.8073
24	0.6848	1.3178	1.7109	2.0639	2.4922	2.7969
25	0.6844	1.3163	1.7081	2.0595	2.4851	2.7874
26	0.6840	1.3150	1.7056	2.0555	2.4786	2.7787

TABLE C.2 317

Degrees of Freedom	Upper-Tail Areas					
	0.25	0.10	0.05	0.025	0.01	0.005
27	0.6837	1.3137	1.7033	2.0518	2.4727	2.7707
28	0.6834	1.3125	1.7011	2.0484	2.4671	2.7633
29	0.6830	1.3114	1.6991	2.0452	2.4620	2.7564
30	0.6828	1.3104	1.6973	2.0423	2.4573	2.7500
31	0.6825	1.3095	1.6955	2.0395	2.4528	2.7440
32	0.6822	1.3086	1.6939	2.0369	2.4487	2.7385
33	0.6820	1.3077	1.6924	2.0345	2.4448	2.7333
34	0.6818	1.3070	1.6909	2.0322	2.4411	2.7284
35	0.6816	1.3062	1.6896	2.0301	2.4377	2.7238
36	0.6814	1.3055	1.6883	2.0281	2.4345	2.7195
37	0.6812	1.3049	1.6871	2.0262	2.4314	2.7154
38	0.6810	1.3042	1.6860	2.0244	2.4286	2.7116
39	0.6808	1.3036	1.6849	2.0227	2.4258	2.7079
40	0.6807	1.3031	1.6839	2.0211	2.4233	2.7045
41	0.6805	1.3025	1.6829	2.0195	2.4208	2.7012
42	0.6804	1.3020	1.6820	2.0181	2.4185	2.6981
43	0.6802	1.3016	1.6811	2.0167	2.4163	2.6951
44	0.6801	1.3011	1.6802	2.0154	2.4141	2.6923
45	0.6800	1.3006	1.6794	2.0141	2.4121	2.6896
46	0.6799	1.3022	1.6787	2.0129	2.4102	2.6870
47	0.6797	1.2998	1.6779	2.0117	2.4083	2.6846
48	0.6796	1.2994	1.6772	2.0106	2.4066	2.6822
49	0.6795	1.2991	1.6766	2.0096	2.4049	2.6800
50	0.6794	1.2987	1.6759	2.0086	2.4033	2.6778
51	0.6793	1.2984	1.6753	2.0076	2.4017	2.6757

(continues)

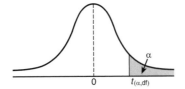

TABLE C.2 Continued

Degrees of Freedom	Upper-Tail Areas					
	0.25	0.10	0.05	0.025	0.01	0.005
52	0.6792	1.2980	1.6747	2.0066	2.4002	2.6737
53	0.6791	1.2977	1.6741	2.0057	2.3988	2.6718
54	0.6791	1.2974	1.6736	2.0049	2.3974	2.6700
55	0.6790	1.2971	1.6730	2.0040	2.3961	2.6682
56	0.6789	1.2969	1.6725	2.0032	2.3948	2.6665
57	0.6788	1.2966	1.6720	2.0025	2.3936	2.6649
58	0.6787	1.2963	1.6716	2.0017	2.3924	2.6633
59	0.6787	1.2961	1.6711	2.0010	2.3912	2.6618
60	0.6786	1.2958	1.6706	2.0003	2.3901	2.6603
61	0.6785	1.2956	1.6702	1.9996	2.3890	2.6589
62	0.6785	1.2954	1.6698	1.9990	2.3880	2.6575
63	0.6784	1.2951	1.6694	1.9983	2.3870	2.6561
64	0.6783	1.2949	1.6690	1.9977	2.3860	2.6549
65	0.6783	1.2947	1.6686	1.9971	2.3851	2.6536
66	0.6782	1.2945	1.6683	1.9966	2.3842	2.6524
67	0.6782	1.2943	1.6679	1.9960	2.3833	2.6512
68	0.6781	1.2941	1.6676	1.9955	2.3824	2.6501
69	0.6781	1.2939	1.6672	1.9949	2.3816	2.6490
70	0.6780	1.2938	1.6669	1.9944	2.3808	2.6479
71	0.6780	1.2936	1.6666	1.9939	2.3800	2.6469
72	0.6779	1.2934	1.6663	1.9935	2.3793	2.6459
73	0.6779	1.2933	1.6660	1.9930	2.3785	2.6449
74	0.6778	1.2931	1.6657	1.9925	2.3778	2.6439
75	0.6778	1.2929	1.6654	1.9921	2.3771	2.6430
76	0.6777	1.2928	1.6652	1.9917	2.3764	2.6421
77	0.6777	1.2926	1.6649	1.9913	2.3758	2.6412

TABLE C.2   319

Degrees of Freedom	Upper-Tail Areas					
	0.25	0.10	0.05	0.025	0.01	0.005
78	0.6776	1.2925	1.6646	1.9908	2.3751	2.6403
79	0.6776	1.2924	1.6644	1.9905	2.3745	2.6395
80	0.6776	1.2922	1.6641	1.9901	2.3739	2.6387
81	0.6775	1.2921	1.6639	1.9897	2.3733	2.6379
82	0.6775	1.2920	1.6636	1.9893	2.3727	2.6371
83	0.6775	1.2918	1.6634	1.9890	2.3721	2.6364
84	0.6774	1.2917	1.6632	1.9886	2.3716	2.6356
85	0.6774	1.2916	1.6630	1.9883	2.3710	2.6349
86	0.6774	1.2915	1.6628	1.9879	2.3705	2.6342
87	0.6773	1.2914	1.6626	1.9876	2.3700	2.6335
88	0.6773	1.2912	1.6624	1.9873	2.3695	2.6329
89	0.6773	1.2911	1.6622	1.9870	2.3690	2.6322
90	0.6772	1.2910	1.6620	1.9867	2.3685	2.6316
91	0.6772	1.2909	1.6618	1.9864	2.3680	2.6309
92	0.6772	1.2908	1.6616	1.9861	2.3676	2.6303
93	0.6771	1.2907	1.6614	1.9858	2.3671	2.6297
94	0.6771	1.2906	1.6612	1.9855	2.3667	2.6291
95	0.6771	1.2905	1.6611	1.9853	2.3662	2.6286
96	0.6771	1.2904	1.6609	1.9850	2.3658	2.6280
97	0.6770	1.2903	1.6607	1.9847	2.3654	2.6275
98	0.6770	1.2902	1.6606	1.9845	2.3650	2.6269
99	0.6770	1.2902	1.6604	1.9842	2.3646	2.6264
100	0.6770	1.2901	1.6602	1.9840	2.3642	2.6259
110	0.6767	1.2893	1.6588	1.9818	2.3607	2.6213
120	0.6765	1.2886	1.6577	1.9799	2.3578	2.6174
∞	0.6745	1.2816	1.6449	1.9600	2.3263	2.5758

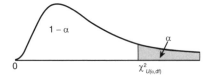

TABLE C.3   Critical Values of $\chi 2$

For a particular number of degrees of freedom, entry represents the critical value of χ^2 corresponding to a specified upper-tail area (α).

Degrees of Freedom	Upper-Tail Areas (α)					
	0.995	0.99	0.975	0.95	0.90	0.75
1			0.001	0.004	0.016	0.102
2	0.010	0.020	0.051	0.103	0.211	0.575
3	0.072	0.115	0.216	0.352	0.584	1.213
4	0.207	0.297	0.484	0.711	1.064	1.923
5	0.412	0.554	0.831	1.145	1.610	2.675
6	0.676	0.872	1.237	1.635	2.204	3.455
7	0.989	1.239	1.690	2.167	2.833	4.255
8	1.344	1.646	2.180	2.733	3.490	5.071
9	1.735	2.088	2.700	3.325	4.168	5.899
10	2.156	2.558	3.247	3.940	4.865	6.737
11	2.603	3.053	3.816	4.575	5.578	7.584
12	3.074	3.571	4.404	5.226	6.304	8.438
13	3.565	4.107	5.009	5.892	7.042	9.299
14	4.075	4.660	5.629	6.571	7.790	10.165
15	4.601	5.229	6.262	7.261	8.547	11.037
16	5.142	5.812	6.908	7.962	9.312	11.912
17	5.697	6.408	7.564	8.672	10.085	12.792
18	6.265	7.015	8.231	9.390	10.865	13.675
19	6.844	7.633	8.907	10.117	11.651	14.562
20	7.434	8.260	9.591	10.851	12.443	15.452
21	8.034	8.897	10.283	11.591	13.240	16.344
22	8.643	9.542	10.982	12.338	14.042	17.240
23	9.260	10.196	11.689	13.091	14.848	18.137
24	9.886	10.856	12.401	13.848	15.659	19.037
25	10.520	11.524	13.120	14.611	16.473	19.939
26	11.160	12.198	13.844	15.379	17.292	20.843
27	11.808	12.879	14.573	16.151	18.114	21.749
28	12.461	13.565	15.308	16.928	18.939	22.657
29	13.121	14.257	16.047	17.708	19.768	23.567
30	13.787	14.954	16.791	18.493	20.599	24.478

For larger values of degrees of freedom (df), the expression $Z = \sqrt{2\chi^2} - \sqrt{2(df) - 1}$ may be used and the resulting upper-tail area can be obtained from the cumulative standardized normal distribution (Table C.1).

TABLE C.3   321

		Upper-Tail Areas (α)			
0.25	**0.10**	**0.05**	**0.025**	**0.01**	**0.005**
1.323	2.706	3.841	5.024	6.635	7.879
2.773	4.605	5.991	7.378	9.210	10.597
4.108	6.251	7.815	9.348	11.345	12.838
5.385	7.779	9.488	11.143	13.277	14.860
6.626	9.236	11.071	12.833	15.086	16.750
7.841	10.645	12.592	14.449	16.812	18.458
9.037	12.017	14.067	16.013	18.475	20.278
10.219	13.362	15.507	17.535	20.090	21.955
11.389	14.684	16.919	19.023	21.666	23.589
12.549	15.987	18.307	20.483	23.209	25.188
13.701	17.275	19.675	21.920	24.725	26.757
14.845	18.549	21.026	23.337	26.217	28.299
15.984	19.812	22.362	24.736	27.688	29.819
17.117	21.064	23.685	26.119	29.141	31.319
18.245	22.307	24.996	27.488	30.578	32.801
19.369	23.542	26.296	28.845	32.000	34.267
20.489	24.769	27.587	30.191	33.409	35.718
21.605	25.989	28.869	31.526	34.805	37.156
22.718	27.204	30.144	32.852	36.191	38.582
23.828	28.412	31.410	34.170	37.566	39.997
24.935	29.615	32.671	35.479	38.932	41.401
26.039	30.813	33.924	36.781	40.289	42.796
27.141	32.007	35.172	38.076	41.638	44.181
28.241	33.196	36.415	39.364	42.980	45.559
29.339	34.382	37.652	40.646	44.314	46.928
30.435	35.563	38.885	41.923	45.642	48.290
31.528	36.741	40.113	43.194	46.963	49.645
32.620	37.916	41.337	44.461	48.278	50.993
33.711	39.087	42.557	45.722	49.588	52.336
34.800	40.256	43.773	46.979	50.892	53.672

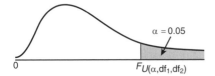

$\alpha = 0.05$

$F_{U(\alpha, df_1, df_2)}$

TABLE C.4   Critical Values of F

For a particular combination of numerator and denominator degrees of freedom, entry represents the critical values of F corresponding to a specified upper-tail area (α).

Denominator,				Numerator, df_1					
df_2	1	2	3	4	5	6	7	8	9
1	161.40	199.50	215.70	224.60	230.20	234.00	236.80	238.90	240.50
2	18.51	19.00	19.16	19.25	19.30	19.33	19.35	19.37	19.38
3	10.13	9.55	9.28	9.12	9.01	8.94	8.89	8.85	8.81
4	7.71	6.94	6.59	6.39	6.26	6.16	6.09	6.04	6.00
5	6.61	5.79	5.41	5.19	5.05	4.95	4.88	4.82	4.77
6	5.99	5.14	4.76	4.53	4.39	4.28	4.21	4.15	4.10
7	5.59	4.74	4.35	4.12	3.97	3.87	3.79	3.73	3.68
8	5.32	4.46	4.07	3.84	3.69	3.58	3.50	3.44	3.39
9	5.12	4.26	3.86	3.63	3.48	3.37	3.29	3.23	3.18
10	4.96	4.10	3.71	3.48	3.33	3.22	3.14	3.07	3.02
11	4.84	3.98	3.59	3.36	3.20	3.09	3.01	2.95	2.90
12	4.75	3.89	3.49	3.26	3.11	3.00	2.91	2.85	2.80
13	4.67	3.81	3.41	3.18	3.03	2.92	2.83	2.77	2.71
14	4.60	3.74	3.34	3.11	2.96	2.85	2.76	2.70	2.65
15	4.54	3.68	3.29	3.06	2.90	2.79	2.71	2.64	2.59
16	4.49	3.63	3.24	3.01	2.85	2.74	2.66	2.59	2.54
17	4.45	3.59	3.20	2.96	2.81	2.70	2.61	2.55	2.49
18	4.41	3.55	3.16	2.93	2.77	2.66	2.58	2.51	2.46
19	4.38	3.52	3.13	2.90	2.74	2.63	2.54	2.48	2.42

TABLE C.4   323

			Numerator, df$_1$						
10	**12**	**15**	**20**	**24**	**30**	**40**	**60**	**120**	**∞**
241.90	243.90	245.90	248.00	249.10	250.10	251.10	252.20	253.30	254.30
19.40	19.41	19.43	19.45	19.45	19.46	19.47	19.48	19.49	19.50
8.79	8.74	8.70	8.66	8.64	8.62	8.59	8.57	8.55	8.53
5.96	5.91	5.86	5.80	5.77	5.75	5.72	5.69	5.66	5.63
4.74	4.68	4.62	4.56	4.53	4.50	4.46	4.43	4.40	4.36
4.06	4.00	3.94	3.87	3.84	3.81	3.77	3.74	3.70	3.67
3.64	3.57	3.51	3.44	3.41	3.38	3.34	3.30	3.27	3.23
3.35	3.28	3.22	3.15	3.12	3.08	3.04	3.01	2.97	2.93
3.14	3.07	3.01	2.94	2.90	2.86	2.83	2.79	2.75	2.71
2.98	2.91	2.85	2.77	2.74	2.70	2.66	2.62	2.58	2.54
2.85	2.79	2.72	2.65	2.61	2.57	2.53	2.49	2.45	2.40
2.75	2.69	2.62	2.54	2.51	2.47	2.43	2.38	2.34	2.30
2.67	2.60	2.53	2.46	2.42	2.38	2.34	2.30	2.25	2.21
2.60	2.53	2.46	2.39	2.35	2.31	2.27	2.22	2.18	2.13
2.54	2.48	2.40	2.33	2.29	2.25	2.20	2.16	2.11	2.07
2.49	2.42	2.35	2.28	2.24	2.19	2.15	2.11	2.06	2.01
2.45	2.38	2.31	2.23	2.19	2.15	2.10	2.06	2.01	1.96
2.41	2.34	2.27	2.19	2.15	2.11	2.06	2.02	1.97	1.92
2.38	2.31	2.23	2.16	2.11	2.07	2.03	1.98	1.93	1.88

(*continues*)

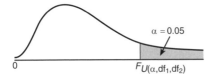

$\alpha = 0.05$

$F_{U(\alpha, df_1, df_2)}$

TABLE C.4   Continued

For a particular combination of numerator and denominator degrees of freedom, entry represents the critical values of F corresponding to a specified upper-tail area (α).

Denominator, df_2	1	2	3	4	5	6	7	8	9
20	4.35	3.49	3.10	2.87	2.71	2.60	2.51	2.45	2.39
21	4.32	3.47	3.07	2.84	2.68	2.57	2.49	2.42	2.37
22	4.30	3.44	3.05	2.82	2.66	2.55	2.46	2.40	2.34
23	4.28	3.42	3.03	2.80	2.64	2.53	2.44	2.37	2.32
24	4.26	3.40	3.01	2.78	2.62	2.51	2.42	2.36	2.30
25	4.24	3.39	2.99	2.76	2.60	2.49	2.40	2.34	2.28
26	4.23	3.37	2.98	2.74	2.59	2.47	2.39	2.32	2.27
27	4.21	3.35	2.96	2.73	2.57	2.46	2.37	2.31	2.25
28	4.20	3.34	2.95	2.71	2.56	2.45	2.36	2.29	2.24
29	4.18	3.33	2.93	2.70	2.55	2.43	2.35	2.28	2.22
30	4.17	3.32	2.92	2.69	2.53	2.42	2.33	2.27	2.21
40	4.08	3.23	2.84	2.61	2.45	2.34	2.25	2.18	2.12
60	4.00	3.15	2.76	2.53	2.37	2.25	2.17	2.10	2.04
120	3.92	3.07	2.68	2.45	2.29	2.17	2.09	2.02	1.96
∞	3.84	3.00	2.60	2.37	2.21	2.10	2.01	1.94	1.88

Numerator, df_1

TABLE C.4   325

				Numerator, df_1					
10	12	15	20	24	30	40	60	120	∞
2.35	2.28	2.20	2.12	2.08	2.04	1.99	1.95	1.90	1.84
2.32	2.25	2.18	2.10	2.05	2.01	1.96	1.92	1.87	1.81
2.30	2.23	2.15	2.07	2.03	1.98	1.91	1.89	1.84	1.78
2.27	2.20	2.13	2.05	2.01	1.96	1.91	1.86	1.81	1.76
2.25	2.18	2.11	2.03	1.98	1.94	1.89	1.84	1.79	1.73
2.24	2.16	2.09	2.01	1.96	1.92	1.87	1.82	1.77	1.71
2.22	2.15	2.07	1.99	1.95	1.90	1.85	1.80	1.75	1.69
2.20	2.13	2.06	1.97	1.93	1.88	1.84	1.79	1.73	1.67
2.19	2.12	2.04	1.96	1.91	1.87	1.82	1.77	1.71	1.65
2.18	2.10	2.03	1.94	1.90	1.85	1.81	1.75	1.70	1.64
2.16	2.09	2.01	1.93	1.89	1.84	1.79	1.74	1.68	1.62
2.08	2.00	1.92	1.84	1.79	1.74	1.69	1.64	1.58	1.51
1.99	1.92	1.84	1.75	1.70	1.65	1.59	1.53	1.47	1.39
1.91	1.83	1.75	1.66	1.61	1.55	1.50	1.43	1.35	1.25
1.83	1.75	1.67	1.57	1.52	1.46	1.39	1.32	1.22	1.00

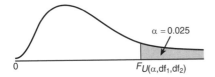

$\alpha = 0.025$

$F_{U(\alpha, df_1, df_2)}$

TABLE C.4   Continued

For a particular combination of numerator and denominator degrees of freedom, entry represents the critical values of F corresponding to a specified upper-tail area (α).

Denominator, df_2	Numerator, df_1								
	1	2	3	4	5	6	7	8	9
1	647.80	799.50	864.20	899.60	921.80	937.10	948.20	956.70	963.30
2	38.51	39.00	39.17	39.25	39.30	39.33	39.36	39.39	39.39
3	17.44	16.04	15.44	15.10	14.88	14.73	14.62	14.54	14.47
4	12.22	10.65	9.98	9.60	9.36	9.20	9.07	8.98	8.90
5	10.01	8.43	7.76	7.39	7.15	6.98	6.85	6.76	6.68
6	8.81	7.26	6.60	6.23	5.99	5.82	5.70	5.60	5.52
7	8.07	6.54	5.89	5.52	5.29	5.12	4.99	4.90	4.82
8	7.57	6.06	5.42	5.05	4.82	4.65	4.53	4.43	4.36
9	7.21	5.71	5.08	4.72	4.48	4.32	4.20	4.10	4.03
10	6.94	5.46	4.83	4.47	4.24	4.07	3.95	3.85	3.78
11	6.72	5.26	4.63	4.28	4.04	3.88	3.76	3.66	3.59
12	6.55	5.10	4.47	4.12	3.89	3.73	3.61	3.51	3.44
13	6.41	4.97	4.35	4.00	3.77	3.60	3.48	3.39	3.31
14	6.30	4.86	4.24	3.89	3.66	3.50	3.38	3.29	3.21
15	6.20	4.77	4.15	3.80	3.58	3.41	3.29	3.20	3.12
16	6.12	4.69	4.08	3.73	3.50	3.34	3.22	3.12	3.05
17	6.04	4.62	4.01	3.66	3.44	3.28	3.16	3.06	2.98
18	5.98	4.56	3.95	3.61	3.38	3.22	3.10	3.01	2.93
19	5.92	4.51	3.90	3.56	3.33	3.17	3.05	2.96	2.88
20	5.87	4.46	3.86	3.51	3.29	3.13	3.01	2.91	2.84
21	5.83	4.42	3.82	3.48	3.25	3.09	2.97	2.87	2.80

TABLE C.4   327

				Numerator, df_1					
10	12	15	20	24	30	40	60	120	∞
968.60	976.70	984.90	993.10	997.20	1,001.00	1,006.00	1,010.00	1,014.00	1,018.00
39.40	39.41	39.43	39.45	39.46	39.46	39.47	39.48	39.49	39.50
14.42	14.34	14.25	14.17	14.12	14.08	14.04	13.99	13.95	13.90
8.84	8.75	8.66	8.56	8.51	8.46	8.41	8.36	8.31	8.26
6.62	6.52	6.43	6.33	6.28	6.23	6.18	6.12	6.07	6.02
5.46	5.37	5.27	5.17	5.12	5.07	5.01	4.96	4.90	4.85
4.76	4.67	4.57	4.47	4.42	4.36	4.31	4.25	4.20	4.14
4.30	4.20	4.10	4.00	3.95	3.89	3.84	3.78	3.73	3.67
3.96	3.87	3.77	3.67	3.61	3.56	3.51	3.45	3.39	3.33
3.72	3.62	3.52	3.42	3.37	3.31	3.26	3.20	3.14	3.08
3.53	3.43	3.33	3.23	3.17	3.12	3.06	3.00	2.94	2.88
3.37	3.28	3.18	3.07	3.02	2.96	2.91	2.85	2.79	2.72
3.25	3.15	3.05	2.95	2.89	2.84	2.78	2.72	2.66	2.60
3.15	3.05	2.95	2.84	2.79	2.73	2.67	2.61	2.55	2.49
3.06	2.96	2.86	2.76	2.70	2.64	2.59	2.52	2.46	2.40
2.99	2.89	2.79	2.68	2.63	2.57	2.51	2.45	2.38	2.32
2.92	2.82	2.72	2.62	2.56	2.50	2.44	2.38	2.32	2.25
2.87	2.77	2.67	2.56	2.50	2.44	2.38	2.32	2.26	2.19
2.82	2.72	2.62	2.51	2.45	2.39	2.33	2.27	2.20	2.13
2.77	2.68	2.57	2.46	2.41	2.35	2.29	2.22	2.16	2.09
2.73	2.64	2.53	2.42	2.37	2.31	2.25	2.18	2.11	2.04

(continues)

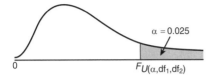

$\alpha = 0.025$

$F_{U(\alpha, df_1, df_2)}$

TABLE C.4   Continued

For a particular combination of numerator and denominator degrees of freedom, entry represents the critical values of F corresponding to a specified upper-tail area (α).

Denominator, df_2	**Numerator, df_1**								
	1	2	3	4	5	6	7	8	9
22	5.79	4.38	3.78	3.44	3.22	3.05	2.93	2.84	2.76
23	5.75	4.35	3.75	3.41	3.18	3.02	2.90	2.81	2.73
24	5.72	4.32	3.72	3.38	3.15	2.99	2.87	2.78	2.70
25	5.69	4.29	3.69	3.35	3.13	2.97	2.85	2.75	2.68
26	5.66	4.27	3.67	3.33	3.10	2.94	2.82	2.73	2.65
27	5.63	4.24	3.65	3.31	3.08	2.92	2.80	2.71	2.63
28	5.61	4.22	3.63	3.29	3.06	2.90	2.78	2.69	2.61
29	5.59	4.20	3.61	3.27	3.04	2.88	2.76	2.67	2.59
30	5.57	4.18	3.59	3.25	3.03	2.87	2.75	2.65	2.57
40	5.42	4.05	3.46	3.13	2.90	2.74	2.62	2.53	2.45
60	5.29	3.93	3.34	3.01	2.79	2.63	2.51	2.41	2.33
120	5.15	3.80	3.23	2.89	2.67	2.52	2.39	2.30	2.22
∞	5.02	3.69	3.12	2.79	2.57	2.41	2.29	2.19	2.11

TABLE C.4   329

				Numerator, df$_1$					
10	**12**	**15**	**20**	**24**	**30**	**40**	**60**	**120**	**∞**
2.70	2.60	2.50	2.39	2.33	2.27	2.21	2.14	2.08	2.00
2.67	2.57	2.47	2.36	2.30	2.24	2.18	2.11	2.04	1.97
2.64	2.54	2.44	2.33	2.27	2.21	2.15	2.08	2.01	1.94
2.61	2.51	2.41	2.30	2.24	2.18	2.12	2.05	1.98	1.91
2.59	2.49	2.39	2.28	2.22	2.16	2.09	2.03	1.95	1.88
2.57	2.47	2.36	2.25	2.19	2.13	2.07	2.00	1.93	1.85
2.55	2.45	2.34	2.23	2.17	2.11	2.05	1.98	1.91	1.83
2.53	2.43	2.32	2.21	2.15	2.09	2.03	1.96	1.89	1.81
2.51	2.41	2.31	2.20	2.14	2.07	2.01	1.94	1.87	1.79
2.39	2.29	2.18	2.07	2.01	1.94	1.88	1.80	1.72	1.64
2.27	2.17	2.06	1.94	1.88	1.82	1.74	1.67	1.58	1.48
2.16	2.05	1.94	1.82	1.76	1.69	1.61	1.53	1.43	1.31
2.05	1.94	1.83	1.71	1.64	1.57	1.48	1.39	1.27	1.00

TABLE C.4   Continued

For a particular combination of numerator and denominator degrees of freedom, entry represents the critical values of F corresponding to a specified upper-tail area (α)

					Numerator, df_1				
Denominator, df_2	**1**	**2**	**3**	**4**	**5**	**6**	**7**	**8**	**9**
1	4,052.00	4,999.50	5,403.00	5,625.00	5,764.00	5,859.00	5,928.00	5,982.00	6,022.00
2	98.50	99.00	99.17	99.25	99.30	99.33	99.36	99.37	99.39
3	34.12	30.82	29.46	28.71	28.24	27.91	27.67	27.49	27.35
4	21.20	18.00	16.69	15.98	15.52	15.21	14.98	14.80	14.66
5	16.26	13.27	12.06	11.39	10.97	10.67	10.46	10.29	10.16
6	13.75	10.92	9.78	9.15	8.75	8.47	8.26	8.10	7.98
7	12.25	9.55	8.45	7.85	7.46	7.19	6.99	6.84	6.72
8	11.26	8.65	7.59	7.01	6.63	6.37	6.18	6.03	5.91
9	10.56	8.02	6.99	6.42	6.06	5.80	5.61	5.47	5.35
10	10.04	7.56	6.55	5.99	5.64	5.39	5.20	5.06	4.94
11	9.65	7.21	6.22	5.67	5.32	5.07	4.89	4.74	4.63
12	9.33	6.93	5.95	5.41	5.06	4.82	4.64	4.50	4.39
13	9.07	6.70	5.74	5.21	4.86	4.62	4.44	4.30	4.19
14	8.86	6.51	5.56	5.04	4.69	4.46	4.28	4.14	4.03
15	8.68	6.36	5.42	4.89	4.56	4.32	4.14	4.00	3.89
16	8.53	6.23	5.29	4.77	4.44	4.20	4.03	3.89	3.78
17	8.40	6.11	5.18	4.67	4.34	4.10	3.93	3.79	3.68
18	8.29	6.01	5.09	4.58	4.25	4.01	3.84	3.71	3.60
19	8.18	5.93	5.01	4.50	4.17	3.94	3.77	3.63	3.52
20	8.10	5.85	4.94	4.43	4.10	3.87	3.70	3.56	3.46
21	8.02	5.78	4.87	4.37	4.04	3.81	3.64	3.51	3.40

TABLE C.4   331

				Numerator, df$_1$					
10	**12**	**15**	**20**	**24**	**30**	**40**	**60**	**120**	**∞**
6,056.00	6,106.00	6,157.00	6,209.00	6,235.00	6,261.00	6,287.00	6,313.00	6,339.00	6,366.00
99.40	99.42	99.43	94.45	99.46	99.47	99.47	99.48	99.49	99.50
27.23	27.05	26.87	26.69	26.60	26.50	26.41	26.32	26.22	26.13
14.55	14.37	14.20	14.02	13.93	13.84	13.75	13.65	13.56	13.46
10.05	9.89	9.72	9.55	9.47	9.38	9.29	9.20	9.11	9.02
7.87	7.72	7.56	7.40	7.31	7.23	7.14	7.06	6.97	6.88
6.62	6.47	6.31	6.16	6.07	5.99	5.91	5.82	5.74	5.65
5.81	5.67	5.52	5.36	5.28	5.20	5.12	5.03	4.95	4.86
5.26	5.11	4.96	4.81	4.73	4.65	4.57	4.48	4.40	4.31
4.85	4.71	4.56	4.41	4.33	4.25	4.17	4.08	4.00	3.91
4.54	4.40	4.25	4.10	4.02	3.94	3.86	3.78	3.69	3.60
4.30	4.16	4.01	3.86	3.78	3.70	3.62	3.54	3.45	3.36
4.10	3.96	3.82	3.66	3.59	3.51	3.43	3.34	3.25	3.17
3.94	3.80	3.66	3.51	3.43	3.35	3.27	3.18	3.09	3.00
3.80	3.67	3.52	3.37	3.29	3.21	3.13	3.05	2.96	2.87
3.69	3.55	3.41	3.26	3.18	3.10	3.02	2.93	2.81	2.75
3.59	3.46	3.31	3.16	3.08	3.00	2.92	2.83	2.75	2.65
3.51	3.37	3.23	3.08	3.00	2.92	2.84	2.75	2.66	2.57
3.43	3.30	3.15	3.00	2.92	2.84	2.76	2.67	2.58	2.49
3.37	3.23	3.09	2.94	2.86	2.78	2.69	2.61	2.52	2.42
3.31	3.17	3.03	2.88	2.80	2.72	2.64	2.55	2.46	2.36

(continues)

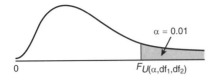

TABLE C.4   Continued

For a particular combination of numerator and denominator degrees of freedom, entry represents the critical values of F corresponding to a specified upper-tail area (α).

Denominator, df$_2$	Numerator, df$_1$								
	1	2	3	4	5	6	7	8	9
22	7.95	5.72	4.82	4.31	3.99	3.76	3.59	3.45	3.35
23	7.88	5.66	4.76	4.26	3.94	3.71	3.54	3.41	3.30
24	7.82	5.61	4.72	4.22	3.90	3.67	3.50	3.36	3.26
25	7.77	5.57	4.68	4.18	3.85	3.63	3.46	3.32	3.22
26	7.72	5.53	4.64	4.14	3.82	3.59	3.42	3.29	3.18
27	7.68	5.49	4.60	4.11	3.78	3.56	3.39	3.26	3.15
28	7.64	5.45	4.57	4.07	3.75	3.53	3.36	3.23	3.12
29	7.60	5.42	4.54	4.04	3.73	3.50	3.33	3.20	3.09
30	7.56	5.39	4.51	4.02	3.70	3.47	3.30	3.17	3.07
40	7.31	5.18	4.31	3.83	3.51	3.29	3.12	2.99	2.89
60	7.08	4.98	4.13	3.65	3.34	3.12	2.95	2.82	2.72
120	6.85	4.79	3.95	3.48	3.17	2.96	2.79	2.66	2.56
∞	6.63	4.61	3.78	3.32	3.02	2.80	2.64	2.51	2.41

TABLE C.4   333

				Numerator, df_1					
10	**12**	**15**	**20**	**24**	**30**	**40**	**60**	**120**	**∞**
3.26	3.12	2.98	2.83	2.75	2.67	2.58	2.50	2.40	2.31
3.21	3.07	2.93	2.78	2.70	2.62	2.54	2.45	2.35	2.26
3.17	3.03	2.89	2.74	2.66	2.58	2.49	2.40	2.31	2.21
3.13	2.99	2.85	2.70	2.62	2.54	2.45	2.36	2.27	2.17
3.09	2.96	2.81	2.66	2.58	2.50	2.42	2.33	2.23	2.13
3.06	2.93	2.78	2.63	2.55	2.47	2.38	2.29	2.20	2.10
3.03	2.90	2.75	2.60	2.52	2.44	2.35	2.26	2.17	2.06
3.00	2.87	2.73	2.57	2.49	2.41	2.33	2.23	2.14	2.03
2.98	2.84	2.70	2.55	2.47	2.39	2.30	2.21	2.11	2.01
2.80	2.66	2.52	2.37	2.29	2.20	2.11	2.02	1.92	1.80
2.63	2.50	2.35	2.20	2.12	2.03	1.94	1.84	1.73	1.60
2.47	2.34	2.19	2.03	1.95	1.86	1.76	1.66	1.53	1.38
2.32	2.18	2.04	1.88	1.79	1.70	1.59	1.47	1.32	1.00

$\alpha = 0.005$

$F_{U(\alpha, df_1, df_2)}$

TABLE C.4 Continued

Denominator, df$_2$	Numerator, df$_1$								
	1	2	3	4	5	6	7	8	9
1	16,211.00	20,000.000	21,615.00	22,500.00	23,056.00	23,437.00	23,715.00	23,925.00	24,091.00
2	198.50	199.00	199.20	199.20	199.30	199.30	199.40	199.40	199.40
3	55.55	49.80	47.47	46.19	45.39	44.84	44.43	44.13	43.88
4	31.33	26.28	24.26	23.15	22.46	21.97	21.62	21.35	21.14
5	22.78	18.31	16.53	15.56	14.94	14.51	14.20	13.96	13.77
6	18.63	14.54	12.92	12.03	11.46	11.07	10.79	10.57	10.39
7	16.24	12.40	10.88	10.05	9.52	9.16	8.89	8.68	8.51
8	14.69	11.04	9.60	8.81	8.30	7.95	7.69	7.50	7.34
9	13.61	10.11	8.72	7.96	7.47	7.13	6.88	6.69	6.54
10	12.83	9.43	8.08	7.34	6.87	6.54	6.30	6.12	5.97
11	12.23	8.91	7.60	6.88	6.42	6.10	5.86	5.68	5.54
12	11.75	8.51	7.23	6.52	6.07	5.76	5.52	5.35	5.20
13	11.37	8.19	6.93	6.23	5.79	5.48	5.25	5.08	4.94
14	11.06	7.92	6.68	6.00	5.56	5.26	5.03	4.86	4.72
15	10.80	7.70	6.48	5.80	5.37	5.07	4.85	4.67	4.54
16	10.58	7.51	6.30	5.64	5.21	4.91	4.69	4.52	4.38
17	10.38	7.35	6.16	5.50	5.07	4.78	4.56	4.39	4.25
18	10.22	7.21	6.03	5.37	4.96	4.66	4.44	4.28	4.14
19	10.07	7.09	5.92	5.27	4.85	4.56	4.34	4.18	4.04
20	9.94	6.99	5.82	5.17	4.76	4.47	4.26	4.09	3.96
21	9.83	6.89	5.73	5.09	4.68	4.39	4.18	4.02	3.88
22	9.73	6.81	5.65	5.02	4.61	4.32	4.11	3.94	3.81
23	9.63	6.73	5.58	4.95	4.54	4.26	4.05	3.88	3.75
24	9.55	6.66	5.52	4.89	4.49	4.20	3.99	3.83	3.69

TABLE C.4   33S

Numerator, df$_1$

10	12	15	20	24	30	40	60	120	∞
24,224.00	24,426.00	24,630.00	24,836.00	24,910.00	25,044.00	25,148.00	25,253.00	25,359.00	25,465.00
199.40	199.40	199.40	199.40	199.50	199.50	199.50	199.50	199.50	199.50
43.69	43.39	43.08	42.78	42.62	42.47	42.31	42.15	41.99	41.83
20.97	20.70	20.44	20.17	20.03	19.89	19.75	19.61	19.47	19.32
13.62	13.38	13.15	12.90	12.78	12.66	12.53	12.40	12.27	12.11
10.25	10.03	9.81	9.59	9.47	9.36	9.24	9.12	9.00	8.88
8.38	8.18	7.97	7.75	7.65	7.53	7.42	7.31	7.19	7.08
7.21	7.01	6.81	6.61	6.50	6.40	6.29	6.18	6.06	5.95
6.42	6.23	6.03	5.83	5.73	5.62	5.52	5.41	5.30	5.19
5.85	5.66	5.47	5.27	5.17	5.07	4.97	4.86	4.75	1.61
5.42	5.24	5.05	4.86	4.75	4.65	4.55	4.44	4.34	4.23
5.09	4.91	4.72	4.53	4.43	4.33	4.23	4.12	4.01	3.90
4.82	4.64	4.46	4.27	4.17	4.07	3.97	3.87	3.76	3.65
4.60	4.43	4.25	4.06	3.96	3.86	3.76	3.66	3.55	3.41
4.42	4.25	4.07	3.88	3.79	3.69	3.58	3.48	3.37	3.26
4.27	4.10	3.92	3.73	3.64	3.54	3.44	3.33	3.22	3.11
4.14	3.97	3.79	3.61	3.51	3.41	3.31	3.21	3.10	2.98
4.03	3.86	3.68	3.50	3.40	3.30	3.20	3.10	2.89	2.87
3.93	3.76	3.59	3.40	3.31	3.21	3.11	3.00	2.89	2.78
3.85	3.68	3.50	3.32	3.22	3.12	3.02	2.92	2.81	2.69
3.77	3.60	3.43	3.24	3.15	3.05	2.95	2.84	2.73	2.61
3.70	3.54	3.36	3.18	3.08	2.98	2.88	2.77	2.66	2.55
3.64	3.47	3.30	3.12	3.02	2.92	2.82	2.71	2.60	2.48
3.59	3.42	3.25	3.06	2.97	2.87	2.77	2.66	2.55	2.43

(continues)

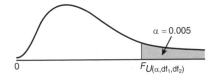

TABLE C.4   Continued

Denominator,	Numerator, df_1								
df_2	1	2	3	4	5	6	7	8	9
25	9.48	6.60	5.46	4.84	4.43	4.15	3.94	3.78	3.64
26	9.41	6.54	5.41	4.79	4.38	4.10	3.89	3.73	3.60
27	9.34	6.49	5.36	4.74	4.34	4.06	3.85	3.69	3.56
28	9.28	6.44	5.32	4.70	4.30	4.02	3.81	3.65	3.52
29	9.23	6.40	5.28	4.66	4.26	3.98	3.77	3.61	3.48
30	9.18	6.35	5.24	4.62	4.23	3.95	3.74	3.58	3.45
40	8.83	6.07	4.98	4.37	3.99	3.71	3.51	3.35	3.22
60	8.49	5.79	4.73	4.14	3.76	3.49	3.29	3.13	3.01
120	8.18	5.54	4.50	3.92	3.55	3.28	3.09	2.93	2.81
∞	7.88	5.30	4.28	3.72	3.35	3.09	2.90	2.74	2.62

TABLE C.4   337

				Numerator, df_1					
10	**12**	**15**	**20**	**24**	**30**	**40**	**60**	**120**	**∞**
3.54	3.37	3.20	3.01	2.92	2.82	2.72	2.61	2.50	2.38
3.49	3.33	3.15	2.97	2.87	2.77	2.67	2.56	2.45	2.33
3.45	3.28	3.11	2.93	2.83	2.73	2.63	2.52	2.41	2.29
3.41	3.25	3.07	2.89	2.79	2.69	2.59	2.48	2.37	2.25
3.38	3.21	3.04	2.86	2.76	2.66	2.56	2.45	2.33	2.21
3.34	3.18	3.01	2.82	2.73	2.63	2.52	2.42	2.30	2.18
3.12	2.95	2.78	2.60	2.50	2.40	2.30	2.18	2.06	1.93
2.90	2.74	2.57	2.39	2.29	2.19	2.08	1.96	1.83	1.69
2.71	2.54	2.37	2.19	2.09	1.98	1.87	1.75	1.61	1.43
2.52	2.36	2.19	2.00	1.90	1.79	1.67	1.53	1.36	1.00

Spreadsheet Tips

The chart and function spreadsheet tips in this appendix complement the Spreadsheet Solutions that appear throughout this book. These tips enable you to learn more about creating or customizing charts and help you better understand the advanced statistical Excel functions that some solutions use. Tips apply equally to Microsoft Windows Excel and Microsoft Excel for Mac, except as noted.

Chart Tips

CT1 Arranging Data in Categorical Charts

The ordering of the data in a worksheet affects the ordering of the data in a visualization. Use the Excel sort command to reorder the rows of data based on a specified condition. In Windows Excel, select **Home → Sort & Filter → Custom Sort**. Then make the appropriate entries in the Sort dialog box and then click **OK**.

For bar charts, reorder the summary table rows from largest value to smallest value. For pie or doughnut charts, use the same reordering in order to show largest to smallest pie slices in clockwise order. If category labels exist in the data to be analyzed, rows of data can be reordered to be in alphabetical order by category using the Excel sort command.

CT2 Reformatting Charts

Excel charts can be reformatted after being created. To reformat a chart, right-click on the chart element to be changed and select the popup menu choice that includes the word **Format**. For example, to change the coloring of bars in a bar or column (vertical bar) chart, right-click a bar and select **Format Data Series**. Selecting the Format choice displays a task pane in which relevant settings can be changed.

To add or remove a chart element, such as an axis label or chart gridlines, select **Chart Design → Add Chart Element**. Then, from the pull-down menu, select the element to add or remove. To change text formatting, select the text and then use items in the **Home** tab as necessary.

CT3 Creating Charts

To create a chart, first select the data to be visualized. (Selection can be done by dragging a mouse over the cell range of the data.) Select **Insert** and click the icon for the chart type desired. Clicking many chart icons, such as the pie chart icon, will display a gallery of choices. Select the appropriate gallery choice to create the chart. To reformat or restyle a chart, use Chart Tip CT2.

CT4 Creating Pareto Charts

To create a Pareto chart, first create a summary table with columns for categories, percentage, and cumulative percentage, ordered from the largest to the smallest percentage value. Then create a combination chart that combines an Excel column chart with a line chart.

To create the chart, first select the ordered summary table. Then select **Insert** and click the **Combo chart icon**. In the combo chart gallery, select the choice identified as **Clustered Column – Line on Secondary Axis**. The chart created will contain two vertical axes—the left axis for the column (bar) chart and the right axis for the line chart—and these axes will, in most cases, have different scales.

To make the combo chart into a proper Pareto chart, rescale each vertical axis to have a maximum value of 1 (100%). To rescale an axis, right-click the axis and select **Format Axis** from the popup menu. In the Format Axis task pane, in the Bounds group under Axis Options, change **Minimum** to **0** and **Maximum** to **1**. Then, reformat the chart as necessary using Chart Tip CT2.

CT5 Creating Histograms

Using the Analysis ToolPak add-in by following the instructions in Advanced Technique ATT1 (see Appendix E) is the simplest method to create a histogram. However, if a frequency distribution has already been created using Advanced Technique ADV2 (see Appendix E), create a new **Column** chart (see Chart Tip CT3) that uses the frequency column of the frequency distribution to visualize the data.

In the column chart that Excel creates, right-click the axis and select **Format Data Series** from the popup menu. In the Format Data Series task pane, change **Gap Width** to **0**. To add midpoint values as labels for the bars, as was done for the Average Ticket Price histogram in Chapter 2, first create a column of midpoint values. Then right-click in the background of the column chart and select **Select Data** from the popup menu. In the Select Data Source dialog box, change **Horizontal (Category) Axis Labels** to the cell range of the midpoint values. (In Excel for Mac, this entry is labeled **Horizontal (Category) axis labels**.) Then click **OK**.

CT6 Creating Time-Series Plots

First, arrange your data values so that the time values to be plotted appear in the column to the immediate left of the values to be plotted on the *Y* axis. Then select **Insert** and click the scatter plot icon. In Windows Excel, this icon is labeled **Insert Scatter (X, Y) or Bubble Chart**. In Excel for Mac, this icon is labeled **X Y (Scatter)**. In either Excel, select the **Scatter with Straight Lines and Markers** choice in the gallery to create the time-series plot.

CT7 Creating Scatter Plots

First, arrange your data values so that the values to be plotted on the (horizontal) *X axis* appear in the column immediately to the left of the values to be plotted on the *Y* axis. Then select **Insert** and click the scatter plot icon. In Windows Excel, this icon is labeled **Insert Scatter (X, Y) or Bubble Chart**. In Excel for Mac, this icon is labeled **X Y (Scatter)**. In either Excel, select the **Scatter** choice in the gallery to create a scatter plot.

Function Tips

FT1 How to Enter Functions for Numerical Descriptive Measures

Enter the functions for numerical descriptive measures in the form *FUNCTION(cell range of the data values)*. Use the AVERAGE (for the mean), MEDIAN, and MODE functions to calculate these measures of central tendency. Use the VAR.S (sample variance), STDEV.S (sample standard deviation), VAR.P (population variance), and STDEV.P (population standard deviation) functions to calculate measures of variation. Use the difference of the MAX (maximum value) and MIN (minimum value) functions to calculate the range.

FT2 Functions for Normal Probabilities

Use the **STANDARDIZE, NORM.DIST, NORM.S.INV**, and **NORM.INV** functions to calculate values associated with normal probabilities. Enter these functions as

- **STANDARDIZE**(X, *mean*, *standard deviation*), where X is the X value of interest, and *mean* and *standard deviation* are the mean and standard deviation for a variable of interest
- **NORM.DIST**(X, *mean*, *standard deviation*, *True*)
- **NORM.S.INV**($P<X$), where $P<X$ is the area under the curve that is less than X
- **NORM.INV**($P<X$, *mean*, *standard deviation*)

The STANDARDIZE function returns the Z value for a particular X value, mean, and standard deviation. The NORM.DIST function returns the area or probability of less than a given X value. The NORM.S.INV function returns the Z value corresponding to the probability of less than a given X. The NORM.INV function returns the X value for a given probability, mean, and standard deviation.

FT3 The FREQUENCY Function

The **FREQUENCY** function creates a frequency distribution from a column of data values using a specified set of *bin values*, which Advanced Technique ADV2 in Appendix E fully explains.

The FREQUENCY function must be entered as part of an *array formula*, a formula that requires a special entry method. To enter this function, first select the cell range that will be containing the frequency counts. Then type, but do not press the **Enter** or **Tab** key, a formula in the form =**FREQUENCY(***cell range of data to be summarized, cell range of the bin values***)**. In Windows Excel, press the **Enter** key while holding down the **Control** and **Shift** keys. In Excel for Mac, press the **Command** key and the **Enter** key as a key combination.

In both Excels, the formula is entered into all cells of the selected cell range, and the formula bar displays curly braces around the formula to indicate that the formula is an array formula. To edit (or delete) an array formula, you must first select all the cells, make the edit, and then press **Enter** while holding down **Control** and **Shift** keys (or, on a Mac, press the **Command** key and the **Enter** key) to complete the edit.

E

Advanced Techniques

Advanced How-To Tips

ADV1 Creating Two-Way Tables Using the Excel PivotTable Feature

Excel PivotTables are worksheet areas that act as if you had entered formulas to summarize data. This feature enables you to create two-way tables from unsummarized data arranged in columns. With a PivotTable, you can *drill down*, or look at decreasing levels of summarization down to the unsummarized data itself.

To create a PivotTable, first arrange your data in consecutive columns, using the first row for a variable name label. (This is the data arrangement used by the downloadable Excel files that Appendix F describes.) Open to the worksheet that contains your unsummarized data and:

1. Select the cell range of the data to be summarized by a PivotTable.
2. Select **Insert → PivotTable**.
3. In the Create PivotTable dialog box, verify that **Select a table or range** is selected and that the **Table/Range** cell range is the cell range selected in step 1. Then, select **New Worksheet** as the location for the PivotTable and click **OK**.
4. In the PivotTable Fields task pane, drag the variable name label for the variable that will be the row variable in the PivotTable and drop it in the **Row Labels** box. Drag a second copy of this same variable name label and drop it in the **Σ Values** box. (This second label changes to **Count of *variable name*.**) Then drag the label for the variable that will be the column variable and drop it in the **Columns Labels** box. Below is the

completed PivotTable Fields task pane for the Chapter 2 two-way table example that tallies entrées ordered by gender for guests during the Friday-to-Sunday weekend period.

Figure E.1 shows the completed PivotTable Fields task pane for the Excel PivotTable version of the Section 2.1 entrées ordered two-way table example.

FIGURE E.1

Completed PivotTable Fields task pane for designing the PivotTable that tallies entrée ordered by sex

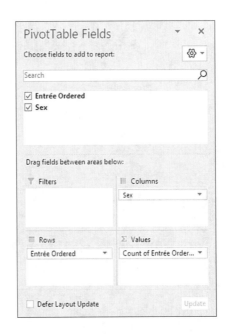

After the PivotTable is created, optionally right-click the PivotTable and select **PivotTable Options** in the popup menu. In the PivotTable Options dialog box, adjust the formatting and display of the PivotTable as necessary and then click **OK**.

ADV2 Creating Frequency Distributions

You use the Microsoft Excel Analysis ToolPak Histogram procedure, which Advanced Technique ATT1 discusses, or the **FREQUENCY** function, which Spreadsheet Tip FT3 in Appendix D describes, to create a frequency distribution.

Using either method, you first create a column of *bin values* on the worksheet that contains the data to be summarized. The bin values define the maximum values of each bin, a grouping of values to be tallied together that is analogous to a frequency distribution group.

When a group is defined in the form *low value through high value*, the bin value is the *high value*. When a group is defined in the form *low value to under high value*, as groups are in Chapter 2, the bin value is a number just smaller than the *high value*. For example, for the group "200 to under 250," the corresponding bin value could be 249.99, a number "just smaller" than 250.

With the column of bin values created, continue by either using the Histogram procedure or the FREQUENCY function. If using the FREQUENCY function, copy the bin values to a new worksheet. Then select the cell range in the next column that is the same size as the cell range of the bin values. Enter the (array) formula that contains the FREQUENCY function into this cell range using the procedure that Spreadsheet Tip FT3 in Appendix D describes.

Figure E.2 shows columns A through C of the Histogram Using FREQUENCY worksheet in the **Chapter 2 Histogram** spreadsheet solution file. The *cell range of data to be summarized* in the **FREQUENCY** function has been entered as Data!B2:B31 and not as B2:B31 because the data values are found on another worksheet—the Data worksheet. Using the name of a worksheet (followed by an exclamation point) as a prefix to a cell range directs Excel to the proper worksheet.

FIGURE E.2

Worksheet formulas found in the Histogram Using FREQUENCY worksheet

	A	B	C
1	Frequency Distribution for Average Ticket Cost		
2	Bins	Frequency	Percentage
3	49.99	=FREQUENCY(Data!B2:B31,A3:A12)	=B3/B$13
4	99.99	=FREQUENCY(Data!B2:B31,A3:A12)	=B4/B$13
5	149.99	=FREQUENCY(Data!B2:B31,A3:A12)	=B5/B$13
6	199.99	=FREQUENCY(Data!B2:B31,A3:A12)	=B6/B$13
7	249.99	=FREQUENCY(Data!B2:B31,A3:A12)	=B7/B$13
8	299.99	=FREQUENCY(Data!B2:B31,A3:A12)	=B8/B$13
9	349.99	=FREQUENCY(Data!B2:B31,A3:A12)	=B9/B$13
10	399.99	=FREQUENCY(Data!B2:B31,A3:A12)	=B10/B$13
11	449.99	=FREQUENCY(Data!B2:B31,A3:A12)	=B11/B$13
12	499.99	=FREQUENCY(Data!B2:B31,A3:A12)	=B12/B$13
13	Total	=SUM(B3:B12)	=SUM(C3:C12)

ADV3 Using the Ampersand Operator to Form Labels

The COMPUTE worksheet of **Chapter 5 Normal** uses formulas in columns A and D to dynamically create labels based on the data values you enter. These formulas make extensive use of the ampersand operator (&) to construct the actual label. For example, the cell A10 formula = "P(X< = "&B8&")" results in the display of **P(X< = 6.2)** because the contents of cell B8, 6.2 are combined with "P(X< = "and")". If you entered the value 9 in cell B8, the label in cell A10 would change to **P(X< = 9)**. Open the Bonus Sheet worksheet of **Chapter 5 Normal** to see all COMPUTE worksheet formulas that use the ampersand operator.

ADV4 Modifying Chapter 6 Sigma Unknown for Use with Unsummarized Data

To modify **Chapter 6 Sigma Unknown** for use with unsummarized data, first enter the unsummarized data in column E. Enter a variable label for the data in cell E1 and then enter values in the cells of the subsequent rows. Do not skip a row.

Next, change the entries in cells B4 through B6. Select cell B4, enter =**STDEV.S(E:E)**, and press **Enter**. Select cell B5, enter =**AVERAGE(E:E)**, and press **Enter**. Select cell B6, enter =**COUNT(E:E)**, and press **Enter**. The worksheet updates all calculations and displays the confidence interval estimate for the unsummarized data.

ADV5 Modifying Chapter 8 Paired T for Use with Other Data Sets

How you modify the **Chapter 8 Paired T** spreadsheet solution for use with another data set depends on the number of values in that other data set. If the data set contains exactly 15 values, enter the data into the worksheet using the instructions found in the worksheet. Otherwise, select cell range E2:H2.

If the data set contains fewer than 15 values, first select the cell range E3:H3. Then follow these steps:

- Right-click over the selected cell range and click **Delete** in the shortcut menu that appears.
- In the Delete dialog box, click **Shift Cells Up** and click **OK**.

Repeat these two bulleted steps as many times as necessary to shorten the data rows to the number of values in the data set. Then enter the new data in columns E through G.

However, if the data contains more than 15 values, follow these steps:

- Right-click over the selected cell range and click **Insert** in the shortcut menu that appears.
- In the Insert dialog box, click **Shift cells down** and click **OK**.

Repeat these two bulleted steps as many times as necessary to lengthen the data rows to the number of values in the data set. Then select cell H2 and copy its formula down to the new blank cells in column H that were just inserted. As a last step, enter the new data in columns E through G.

ADV6 Using the LINEST Function to Calculate Regression Results

You can use the Excel **LINEST** function in an array formula to calculate regression results in Microsoft Excel. These results are similar to those calculated by the Analysis ToolPak Regression procedure.

To use this function, first, with your mouse, select an empty cell range that is five rows deep and contains the number of columns that is equal to the number of your independent X variables plus one. For a simple linear regression model, select five rows by two columns; for a multiple regression model with two independent variables, select five rows by three columns.

Type, but do not press the **Enter** or **Tab** key, a formula in the form =LINEST(*cell range of the dependent variable, cell range of the independent variables*, **True, True**). Then, in Windows Excel, while holding down the **Control** and **Shift** keys, press the **Enter** key. In Excel for Mac, press the **Command** key and the **Enter** key as a key combination.

This enters the formula as an array formula in all the cells you previously selected. To edit (or delete) the array formula, you must first select all the cells, make the edit, and then press **Enter** while holding down the **Control** and **Shift** keys (or press the **Command** key and the **Enter** key) to complete the edit.

The results returned by the array formula are unlabeled. For a multiple regression model, some results appear as **#N/A**, which is not an error. Add labels in the columns immediately to the left and right of the results area to label the results, as shown in the example below.

In **Appendix E Regression**, the **SLR_LINEST** and the **MR_LINEST** worksheets illustrate this labeling technique. Shown below is the MR-LINEST worksheet using the real estate study data from WORKED-OUT PROBLEM 1 in Chapter 11. In this worksheet, the LINEST array function has been entered into the cell range B3 through D7.

	A	B	C	D	E
1	Multiple Regression Analysis for Real Estate Study				
2					
3	Age Coefficient (b_2)	-5.7672	0.0208	930.1169	Property Size Coefficient (b_1), Intercept Coefficient (b_0)
4	Large SS	1.0025	0.0022	84.5721	Property Size *Standard Error*, Intercept *Standard Error*
5	R Square	0.7017	227.2310	#N/A	Standard Error
6	F	69.3806	59.0000	#N/A	Residual *df*
7	Total SS	7164791.1190	3046402.8165	#N/A	Residual *MS*

ADV7 Modifying Chapter 10 Simple Linear Regression for Use with Other Data Sets

To modify **Chapter 10 Simple Linear Regression** for use with other data sets, open to the SLRData worksheet and paste the new independent variable data into column A and the new dependent variable data in column B, replacing the data already there. Then open to the SLR worksheet and select the cell range L2:M6 and edit the cell range in the LINEST function of the array formula (see Spreadsheet Tip FT3 in Appendix D). The worksheet updates the results to reflect the new data set.

Modifying the RESIDUALS worksheet is a multistep process. First, paste the independent variable data into column B and the dependent variable data in column D of the RESIDUALS worksheet. Then, for sample sizes smaller than 36, delete the extra rows. For sample sizes greater than 36, copy the column C and E formulas down through the row containing the last pair of variable values and add the new observation numbers in column A.

ADV8 Creating Excel Bullet Graphs

Because the bullet graph is not an Excel chart type, the **Chapter 12 Bullet Graph** spreadsheet solution plots five different stacked bar charts in one chart to create its bullet graph. These five stacked bar charts correspond to the five data sets labeled as Actual Bar, Poor Range, Acceptable Range, Target Gap, and Target Line. Values in these data sets other than Actual Bar are scaled by formulas that use the highest (maximum) target value and the "poor" and "acceptable" percentages of a target value found in cells L2 and L3 of the BulletData worksheet.

The following figure shows the formulas for the five data series, located in columns E through I of the BulletData worksheet.

FIGURE E.3

Worksheet formulas for the five data series the bullet graph in *Chapter 12 Bullet* Graph spreadsheet solution uses

▲	E	F	G	H	I
1					
2			Calculations		
3	Poor	Acceptable	Acceptable Length	Target Gap	Target Line
4	=$L$2*B4	=$L$3*B4	=F4-E4	=B4-F4	=0.008*MAX(B:B)
5	=$L$2*B5	=$L$3*B5	=$L$3*B5-E5	=B5-F5	=0.008*MAX(B:B)
6	=$L$2*B6	=$L$3*B6	=$L$3*B6-E6	=B6-F6	=0.008*MAX(B:B)
7	=$L$2*B7	=$L$3*B7	=$L$3*B7-E7	=B7-F7	=0.008*MAX(B:B)
8	=$L$2*B8	=$L$3*B8	=$L$3*B8-E8	=B8-F8	=0.008*MAX(B:B)
9	=$L$2*B9	=$L$3*B9	=$L$3*B9-E9	=B9-F9	=0.008*MAX(B:B)

The Poor, Acceptable Length, and Target Gap columns calculate values for three segments colored dark gray, light gray, and white of what appears to be one stacked bar chart in the BulletGraph graph sheet. Target Line uses a small percentage of the maximum value to properly scale the width of the target line at the end the Target Gap segments. A separate bar chart visualized the actual values in column C (not shown above). Bars for this chart are black and have narrower widths than the others.

To create bullet graphs for other data, enter the data in columns A, B, and C and copy down the formulas in columns E through I to the last row of the new data. In the altered chart, select the horizontal (Unit Sale) axis, right-click, and select **Format Axis** from the popup menu. In the Format Axis task pane, in the Bounds group under Axis Options, change **Maximum** to the highest target value.

Analysis ToolPak Tips

The Analysis ToolPak add-in component of Microsoft Excel adds a number of statistical procedures. Use the instructions in Section A.2 to verify that this add-in is installed on your computer. (The Analysis ToolPak is normally installed when you install Microsoft 365 Excel.)

To use an Analysis ToolPak procedure, select **Data → Data Analysis**. In the Data Analysis dialog box, select a procedure from the **Analysis Tools** list and then click the **OK** button. In the dialog box below, **Regression** has been selected for the list.

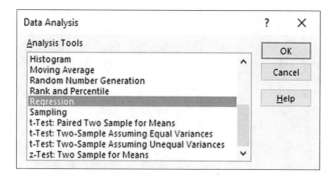

For most procedures, a second dialog box will appear after you click the OK button. In that dialog box, make entries and selections and click **OK** to run the analysis.

ATT1 Histogram Procedure

Begin by making sure that the worksheet that contains your data values also contains a column of bin values (see Advanced Technique ADV2, earlier in this appendix). Select **Data → Data Analysis**. In the Data Analysis dialog box, select **Histogram** and click **OK**.

In the Histogram dialog box, enter the cell ranges of the data values and bin values and check **Labels** if these ranges begin with a column heading. Check **Chart Output** if you want to create a histogram as well as a frequency distribution.

ATT2 Descriptive Statistics Procedure

Place the data values for the variable to be summarized in a column, using the row 1 cells for column labels. Select **Data → Data Analysis**. In the Data Analysis dialog box, select **Descriptive Statistics** and click **OK**.

In the Descriptive Statistics dialog box, enter that column range as the **Input Range**, click **Columns** in the **Grouped By** set, and check **Labels in First Row**. Then click **New Worksheet Ply** and check **Summary statistics**. A table of descriptive statistics appears on a new worksheet.

ATT3 t-Test: Two-Sample Assuming Equal Variances Procedure

Place the data for the two groups in separate columns, using the row 1 cells for column (group) labels. Select **Data → Data Analysis**. In the Data Analysis dialog box, select **t-Test: Two-Sample Assuming Equal Variances** and click **OK**.

In the t-Test: Two-Sample Assuming Equal Variances dialog box, enter the group 1 cell range as the **Variable 1 Range** and enter the group 2 cell range as the **Variable 2 Range**. Enter **0** as the **Hypothesized Mean Difference** and check **Labels**. Enter **0.05** as the **Alpha** value, select the **New Worksheet Ply** option, and click **OK**.

The results will appear on a new worksheet similar to the Figure E.4 worksheet for the data of WORKED-OUT PROBLEM 4 in Chapter 8.

FIGURE E.4

Worksheet results for the two-sample *t*-test assuming equal variances

	A	B	C
1	t-Test: Two-Sample Assuming Equal Variances		
2			
3		*First Half*	*Second Half*
4	Mean	5.7893	5.5345
5	Variance	0.4928	0.5031
6	Observations	28	29
7	Pooled Variance	0.4980	
8	Hypothesized Mean Difference	0	
9	df	55	
10	t Stat	1.3627	
11	P(T<=t) one-tail	0.0893	
12	t Critical one-tail	1.6730	
13	P(T<=t) two-tail	0.1785	
14	t Critical two-tail	2.0040	

ATT4 t-Test: Paired Two Sample for Means Procedure

Place the data for the two groups in separate columns, using the row 1 cells for column (group) labels. Select **Data → Data Analysis** and, in the Data Analysis dialog box, select **t-Test: Paired Two Sample for Means** and click **OK**.

In the t-Test: Paired Two Sample for Means dialog box, enter the group 1 sample data cell range as the **Variable 1 Range** and enter the group 2 sample data cell range as the **Variable 2 Range**. Enter **0** as the **Hypothesized Mean Difference** and check **Labels**. Enter **0.05** as the **Alpha** value, select the **New Worksheet Ply** option, and click **OK**.

The results will appear on a new worksheet similar to the Figure E.5 worksheet for the data of WORKED-OUT PROBLEM 6 in Chapter 8 shown below.

FIGURE E.5

Worksheet results for the paired two-sample *t*-test

	A	B	C
1	t-Test: Paired Two Sample for Means		
2			
3		*Inexpensive Restaurant Meal*	*McDonalds McMeal*
4	Mean	10.2664	7.1676
5	Variance	43.1044	7.2397
6	Observations	25	25
7	Pearson Correlation	0.8027	
8	Hypothesized Mean Difference	0	
9	df	24	
10	t Stat	3.3045	
11	P(T<=t) one-tail	0.0015	
12	t Critical one-tail	1.7109	
13	P(T<=t) two-tail	0.0030	
14	t Critical two-tail	2.0639	

ATT5 ANOVA: Single Factor Procedure

Place the data of each group in its own column, using the row 1 cells for column (group) labels. Select **Data → Data Analysis** and, in the Data Analysis dialog box, select **ANOVA: Single Factor** and click **OK**.

In the ANOVA: Single Factor dialog box, enter the cell range for *all* of your data as the **Input Range**. Select **Columns** and check **Labels in First Row**. Enter **0.05** as the **Alpha** value, select the **New Worksheet Ply** option, and click **OK**.

The results appear on a new worksheet that will be similar to the **Chapter 9 One-Way ANOVA** spreadsheet solution, except that there will be a row that contains the level of significance. **Chapter 9 ATP One-Way ANOVA** shows the Analysis ToolPak results that are similar to the Figure 9.5 results for the Chapter 9 WORKED-OUT PROBLEM 7.

ATT6 Regression Procedure

Place the data for each variable in its own column, using the row 1 cells for column (group) labels. Use the first column for the dependent variable and use the second and subsequent columns for your independent variables. (Simple linear regression, discussed in Chapter 10, uses only one independent variable.) Select **Data → Data Analysis** and, in the Data Analysis dialog box, select **Regression** and click **OK**.

In the Regression dialog box, enter the cell range for dependent variable Y as the **Input Y Range** and enter the cell range for independent X variable or variables as the **Input X Range**. Check **Labels** and **Confidence Level** and enter **95** in the percentage box. To assist in a residual analysis, also check **Residuals** and **Residual Plots**. Click **OK**. Regression results appear on a new worksheet that contains a layout similar to the regression results spreadsheets that Chapters 10 and 11 discuss.

Chapter 10 ATP Simple Linear Regression contains the ToolPak regression results that are similar to columns A through I of the Figure 10.2 worksheet results for the Chapter 10 WORKED-OUT PROBLEM 2 motion picture distributor study data. **Chapter 11 ATP Multiple Regression** contains the ToolPak regression results that are similar to columns A through I of the Figure 11.1 worksheet results for the Chapter 11 WORKED-OUT PROBLEM 1 real estate study data.

F

Documentation for Downloadable Files

This appendix lists and describes all of the files that you can download for use with this book. These files can be found on the webpage for this book, which, at time of publication, exists in the InformIT website at https://informit.com.

F.1 Downloadable Data Files

Throughout this book, the file icon identifies downloadable data files that allow you to examine the data for selected problems. Each data file is available as an Excel .xlsx workbook file.

Aldi Publix Prices	Product, Aldi price, Publix price
Analysis of Sales	Sales type, beverage sales, bakery sales, grocery sales, frozen sales, organic sales, paper goods sales (Chapter 13 JMP example)
Anscombe	Data sets A, B, C, and D—each with 11 pairs of X and Y values
Baseball	Team, win percentage, ERA, runs scored per game
Big Mac Starbucks	City, Big Mac cost, Starbucks tall latte cost
Box Fills	Plant 1 cereal weights, Plant 2 cereal weights, Plant 3 cereal weights, and Plant 4 cereal weights
Cat Food	Ounces eaten of kidney, shrimp, chicken liver, salmon, and beef
Coffee Brew Ratings	Reviewer, Coffeepot, K-Cup
Convenience Store Analysis	Transaction ID, item category

Domestic Beer	Brand, alcohol percentage, carbohydrates, and calories
Flashlight Sales	Checkout sales, end-aisle sales, in-aisle sales
Glen Cove	Address, asking price, property size in acres, age of house, number of bedrooms, number of bathrooms
Insurance	Time to process applications
Intaglio	Surface hardness of untreated steel plates, surface hardness of treated steel plates
Iris Plants	Sepal length, sepal width, petal length, petal width (Chapter 13 JMP example)
Math	Math scores using set A materials, math scores using set B materials, math scores using set C materials
MLB Salaries	Year, average MLB salary
Movie Releases	Year, number of movies released
Movies	Trailer views (millions), opening weekend ($millions)
Moving	Labor hours, cubic feet moved, and number of large pieces of furniture moved
MPG	Auto, Overall MPG
Myeloma	Patient, measurement before transplant, and measurement after transplant
NBA Ticket Cost	Team, average ticket cost ($), premium ticket cost ($)
Order Time Samples	Data for order time samples
Order Time Population	Order time
Potter Movies	Title, first weekend gross ($millions), total domestic gross ($millions)
Protein	Food, calories, protein (grams), percentage calories from fat, percentage calories from saturated fats, cholesterol (mg)
QSR Sales	Mean sales per unit for burger, chicken, sandwich, and pizza market segments
Redwood	Height, breast-height diameter, bark thickness
Restaurants	Location, food rating, decor rating, service rating, summated rating, coded location (0 = city, 1 = suburban), cost of meal ($)
Super Bowl Ads	Brand, average ad score, half shown (also contains other worksheets that reorganize this data)
Supermarket	Pair, sales using new package, sales using old package, difference in sales
Times	Times to get ready
UHDTV Wholesale Sales	Year, wholesale sales ($millions)

| Vintage Autos | MPG, horsepower, weight, acceleration, displacement (Chapter 13 JMP example) |
| World Meal Costs | City, inexpensive restaurant meal (US$), McDonald's McMeal (US$) |

f.2 Downloadable Spreadsheet Solution Files

Also available for download are the Excel workbook files that are mentioned in the *Spreadsheet Solution* sections of this book. The following is a complete list of the Spreadsheet Solution Excel workbook files for this book.

Chapter 2 Bar

Chapter 2 Doughnut

Chapter 2 Histogram

Chapter 2 Pareto

Chapter 2 Pie

Chapter 2 Scatter Plot

Chapter 2 Time-Series

Chapter 2 Two-Way PivotTable

Chapter 2 Two-Way

Chapter 3 Boxplot

Chapter 3 Descriptive

Chapter 3 WorkedOut Problem 7

Chapter 3 WorkedOut Problem 15

Chapter 5 Binomial

Chapter 5 Normal

Chapter 5 Poisson

Chapter 6 Proportion

Chapter 6 Sigma Unknown

Chapter 8 ATP Paired T

Chapter 8 ATP Pooled-Variance T

Chapter 8 Paired T

Chapter 8 Paired T Advanced

Chapter 8 Pooled-Variance T with Sample Statistics

Chapter 8 Pooled-Variance T with Unsummarized Data

Chapter 8 Z Two Proportions

Chapter 9 ATP One-Way ANOVA

Chapter 9 Chi-Square

Chapter 9 Chi-Square Spreadsheets

Chapter 9 One-Way ANOVA

Chapter 9 WorkedOut Problem 9

Chapter 10 ATP Simple Linear Regression

Chapter 10 Simple Linear Regression

Chapter 11 ATP Multiple Regression

Chapter 11 Multiple Regression

Chapter 12 Bullet Graph

Chapter 12 PivotTable

Chapter 12 Sparklines

Chapter 12 Treemap

Appendix E Regression

Index

C

X-Y-Z

Photo by izusek/gettyimages

Register Your Product at informit.com/register

Access additional benefits and **save 35%** on your next purchase

- Automatically receive a coupon for 35% off your next purchase, valid for 30 days. Look for your code in your InformIT cart or the Manage Codes section of your account page.

- Download available product updates.

- Access bonus material if available.*

- Check the box to hear from us and receive exclusive offers on new editions and related products.

Registration benefits vary by product. Benefits will be listed on your account page under Registered Products.

InformIT.com—The Trusted Technology Learning Source

InformIT is the online home of information technology brands at Pearson, the world's foremost education company. At InformIT.com, you can:

- Shop our books, eBooks, software, and video training
- Take advantage of our special offers and promotions (informit.com/promotions)
- Sign up for special offers and content newsletter (informit.com/newsletters)
- Access thousands of free chapters and video lessons

Connect with InformIT—Visit informit.com/community

the trusted technology learning source

Addison-Wesley • Adobe Press • Cisco Press • Microsoft Press • Pearson IT Certification • Prentice Hall • Que • Sams • Peachpit Press

 Pearson

Analytics & Metrics
Books, eBooks & Video

InformIT has a catalog on data analysis and metrics for various industries and level of expertise.

Visit **informit.com/analytics-center** to shop and preview sample chapters on topics including:

- Prescriptive and predictive analytics
- Product analytics
- Digital marketing
- Ecommerce
- Global business
- Advanced methodologies
- Analytics with Excel & Power BI